Ayman Mohamady Atta

Demulsification of Crude Oil Emulsion

Ahmed Ali Fadda
Rasha Refat Fouad
Ayman Mohamady Atta

Demulsification of Crude Oil Emulsion

Utility of Modified Polypropylene oxide-Polyethylene oxide Copolymer as Demulsifier Agents

LAP LAMBERT Academic Publishing

Impressum/Imprint (nur für Deutschland/only for Germany)
Bibliografische Information der Deutschen Nationalbibliothek: Die Deutsche Nationalbibliothek verzeichnet diese Publikation in der Deutschen Nationalbibliografie; detaillierte bibliografische Daten sind im Internet über http://dnb.d-nb.de abrufbar.

Coverbild: www.ingimage.com

Verlag: LAP LAMBERT Academic Publishing GmbH & Co. KG
Heinrich-Böcking-Str. 6-8, 66121 Saarbrücken, Deutschland
Telefon +49 681 3720-310, Telefax +49 681 3720-3109
Email: info@lap-publishing.com

Approved by: Mansoura, Mansoura University, Diss., 2012

Herstellung in Deutschland (siehe letzte Seite)
ISBN: 978-3-659-16087-5

Imprint (only for USA, GB)
Bibliographic information published by the Deutsche Nationalbibliothek: The Deutsche Nationalbibliothek lists this publication in the Deutsche Nationalbibliografie; detailed bibliographic data are available in the Internet at http://dnb.d-nb.de.

Cover image: www.ingimage.com

Publisher: LAP LAMBERT Academic Publishing GmbH & Co. KG
Heinrich-Böcking-Str. 6-8, 66121 Saarbrücken, Germany
Phone +49 681 3720-310, Fax +49 681 3720-3109
Email: info@lap-publishing.com

Printed in the U.S.A.
Printed in the U.K. by (see last page)
ISBN: 978-3-659-16087-5

CONTENTS

ABBEREVIATIONS

W/O	Water-in-oil emulsions
O/W	Oil-in-water emulsions
W/O/W	Water-in-oil-in-water
CMC	Critical micelle concentration
CMT	Critical micelle temperature
PO	Propylene oxide
EO	Ethylene oxide
PEO	Poly (ethylene oxide)
PPO	Poly (propylene oxide)
PEGME	Poly (ethylene glycol) monomethyl ether
PVA	Poly (vinyl alcohol)
Vac	Vinyl acetate
NMR	Nuclear magnetic resonance
HLB	Hydrophil-lipophil balance
MA	Maleic anhydride
DBP	Dibenzoyl peroxide
PTSA	P-toluene sulfonic acid
KOH	Potassium hydroxide
BO	Butylene oxide
PVP	Poly (vinylpyrrolidone)
PMMA	Poly(methyl methacrylate)
PTMO	Poly(tetramethylene oxide)
DBSA	P-(n-dodecyl) benzene sulfonic acid
PEO-PPO-PEO	Poly(ethylene oxide)–poly(propylene oxide)–poly(ethylene oxide)
PEG	Poly ethylene glycol
PPG	Poly propylene glycol
IFT	Interfacial tension
MALDI-TOF-MS	Matrix-assisted laser desorption/ionization time-of-flight mass spectrometry
LCCC	Liquid chromatography under critical conditions
CID	Collision induced dissociation
HTAB	Hexadecyltrimethyl ammonium bromide
TTAB	Tetradecyltrm-ethylammonium bromide
ESCA	Electron spectroscopy for chemical analysis
SPR	Surface plasma resonance
ATRP	Atom transfer radical polymerization
DSC	Differential scanning calorimetry
GPC	Gel permeation chromatography
MW	Molecular weight
HMWHCs	High molecular weight hydrocarbons

AIM OF THE WORK

The formation of water-in-crude oil emulsion causes several economic and environmental problems such as corrosion in petroleum pipelines and catalyst poisonous at the refiners.

Such emulsions are naturally found in reservoirs or they are the result of various extraction or cleaning processes. So for economic and operational reasons, it is necessary to separate the water completely from the crude oils before transporting or refining them. Many studies were carried out for synthesis of polymeric surfactants for treating crude oil emulsions.

Polymeric surfactants are useful as demulsifiers, they can reduce the surface tension of aqueous phase, stabilize polymer particles and can form aggregates as the result of intermolecular and/or intramolecular hydrophobic interactions.

The present work aimed to prepare water soluble non-ionic surfactants from ethylene oxide – propylene oxide copolymers by grafting technique. Poly (ethylene glycol)–block–poly (propylene oxide)–block–poly (ethylene glycol) copolymers, PEG-PPO-PEG, were prepared under normal condition.

Evaluation of the prepared polymeric surfactants, as demulsifiers for asphaltenic crude oil emulsions is the main goal of the present work.

ABSTRACT

Water-in-crude oil emulsions stabilized by natural surface-active components (such as asphaltene) are one of the major problems in relation to petroleum production. New water soluble nonionic amphiphilic block and graft copolymers based on hydrophilic poly (ethylene glycol) (PEG) and hydrophobic poly (propylene oxide) (PPO) were prepared. PEG-PPO-PEG copolymers were prepared in the normal condition by two methods to use them as demulsifiers for synthetic crude oil emulsions. In the first method PPO was esterified with MA to produce poly (propylene oxide) maleate diester. The produced polymers were esterified with PEG having different molecular weights to produce PEG-PPO-PEG block copolymers. The second method was based on grafting of MA onto PPO followed by esterification with PEG. The first step in this work is to polymerize propylene glycol and propylene oxide using KOH as a catalyst to produce poly (propylene glycol), PPO1, at the reaction temperature 298 K. The reaction time was adjusted to produce PPO having different molecular weights. The molecular weights of the prepared PEG-PPO-PEG block copolymers were determined by GPC technique. The structure of the prepared surfactants was illustrated by IR and ^{1}H-NMR spectra. The surface activity of the prepared surfactants was determined from the surface and thermodynamic properties of these surfactants. The evaluation of the prepared poly(oxyethylene)-co-poly(oxypropylene) block and graft copolymers as demulsifiers for water-in-crude oil emulsions was carried out at 318 and 333 K. Data indicated that the prepared demulsifiers showed good demulsification efficiencies towards the different types of water-in-crude oil emulsions.

ACKNOWLEDGEMENT

First of all thanks for **Allah** the most beneficent, the most merciful. I wish to express my sincere appreciation and gratitude to **Prof. Dr. Ahmed Ali Hamed Fadda**, Professor of Organic Chemistry, Faculty of Science, Mansoura University, **Prof. Dr. Adel Abdel Hady Nassar**, Professor of Organic Chemistry, Faculty of Science, Menofia University and **Prof. Dr. Aiman Mohamdy Atta**, Prof. of Polymer Chemistry, Egyptian Petroleum Research Institute for suggesting the research point, their continuous guidance, precious advice, real support and valuable criticism.

I would to express my deep appreciation and gratitude to **Dr. Hussin Al-Shafey**, Prof. of Polymer Chemistry, Petroleum Research Institute for his helping.

I would to express my deep appreciation and gratitude to all my family especially for my **mother**, **brother**, **my son** and **my husband Zakaria**.

Finally, I am really grateful to the staff members of Chemistry Department, Faculty of Science, Mansoura University, for their support.

Rasha Rifat Fouad Orban

CHAPTER I

"INTRODUCTION"

1.1. Introduction

The name amphiphile is sometimes used synonymously with surfactant. The word is derived from the Greek word *amphi*, meaning both and the term relates to the fact that all surfactant molecules consist of at least two parts, one which is soluble in a specific fluid (the lyophilic part) and one which is insoluble (the lyophobic part). The hydrophobic part of a surfactant may be branched or linear. The polar head group is usually but not always attached at one end of the alkyl chain. The length of the chain is in the range of 8-18 carbon atoms. The degree of chain branching, the position of the polar group and the length of the chain are very important parameters for the physicochemical properties of the surfactant. The polar part of the surfactant may be ionic or non-ionic and the choice of polar group determines the properties to a large extent. The relative size of the hydrophobic and polar groups not the absolute size of either of the two is decisive in determining the physicochemical behavior of a surfactant in water. Water-soluble synthetic polymers are a family of materials that have been developed commercially and studied scientifically at an accelerating pace in recent years. Partly, this is a reflection of the increasing diversity in the applications of water-soluble polymers in mineral processing, water-treatment, oil-recovery and surfactants. Surfactants are used together with polymers in a wide range of applications. In areas as diverse as detergents, paints, paper coatings, food and pharmacy, formulations usually contain a combination of a low molecular weight surfactant and a polymer which may or may not be highly surface active. Together, the surfactant and the polymer provide the stability, rheology, etc. needed for specific application.

The formation of water-in-crude oil emulsions is in the centre of several economic and environmental problems [1]. Such emulsions are naturally found in reservoirs or they are the result of various extraction or cleaning processes [2, 3]. Many studies have shown that the stability of water-in-crude oil emulsions depends mainly on rigid protective film encapsulating the water droplets [4]. This rigid interfacial film is believed by many researchers to be composed predominantly of asphaltenes, resins, and/or fine solids [5]. These emulsions attain their stability from the presence of asphaltenes, which are condensed aromatic rings containing saturated carbon chain and naphthenic rings as substituents, along with a distribution of heteroatoms and metals. Asphaltenes are known to stabilize water-in-oil emulsions, making it more diffcult to demulsify and desalt crude oils [6]. In an asphaltene-stabilized emulsion, the asphaltenes are belived to exist as a rigid, cross-linked network absorbed from the oil to the oil-water interface [7]. The high stability of the emulsions is belived to be due to strong interficial films formed by the asphaltene networks [8]. These emulsions create problems during the production and transportation of the multiphase oil–water–gas mixtures to the process plants from the production sites. This increases the cost of production and transportation of petroleum. Environmental problems arise because of the difficulty in cleaning up of the environment after oil spillage by techniques such as burning, use of sorbants, use of dispersants and pumping [1, 9]. So for economic and operational reasons, it is necessary to separate the water completely from the crude oils before transporting or refining them [10, 11].

The problem of resolving water-in-oil emulsions has been attacked in a number of ways over the past several years and a great deal progress has been made in developing and utilizing methods for demulsification. Morrel and Egloff [12] have listed seven basic methods by which emulsion may be denatured: (1) settling, (2) heating or distilling at atmospheric pressure, (3) heating or distilling at elevated pressure, (4) use of demulsifiers, (5) electrical dehydration, (6) centrifuging, and (7) filtration. All of these methods and combinations of these methods are used at present time. However, the chemical demulsification process is by far the most widely used method in petroleum industry [11]; thus demulsification was an effective method to accelerate oil-water separation [13]. Chemical demulsification using demulsifiers is in common use in oil field chemistry. The demulsification ability of a demulsifier is mainly controlled by two factors: one is the hydrophilic–hydrophobic ability; the other is the ability to destroy the interfacial film [14].

The structure of the demulsifier can influence both of above two factors.
The interfacial shear property has been the most investigated viscoelastic property of liquid–liquid interface films [15].

Oil soluble demulsifiers are commonly used to destabilize the W/O emulsions. These demulsifiers are polydisperse, interfacial active polymers and are, mostly, non-ionic block polymers with hydrophilic and hydrophobic segments. The demulsification method involves the addition of the demulsifiers in-situ in concentrations, depending on the total effective area of the dispersed phase. In most cases, the manufacture of these demulsifiers involves the handling of dangerous and expensive chemicals like ethylene and propylene oxide. It would be very desirable to have water soluble demulsifiers that are as effective as their oil soluble counterparts [16].

Commercial demulsifiers are polymeric surfactants such as copolymers of polyoxyethylene and polypropylene or polyester or blends of different surface-active substances [17].Polyoxyethylene-polyoxypropylene block copolymers consist of hydrophobic and hydrophilic parts. They are nonionic surface active agents and in aqueous solutions, form micelles at a certain critical micelle concentration (CMC) [18]. Because of their amphiphilic nature, these copolymers are adsorbed at various interfaces. A great number of adsorption investigations at solid/liquid interfaces are known [19–21]. Experiments show that the character of the interface is significant for the equilibrium adsorption and for the area occupied by one molecule at the interface. The comparison of the experimental data for polyoxyethylene-polyoxypropylene block copolymers leads to the conclusion that the area per molecule is larger in the case of a solution/hydrophilic solid interface[19], smaller in the case of a solution/hydrophobic solid interface [20] and smallest for the air/solution interface [21].

1.2. General Characteristics of Polymeric Surfactants

Surfactant is characterized by its tendency to adsorb at surfaces and interfaces. The surfactant molecule should possess three characteristics, the molecular structure should be composed of polar and non-polar groups, it should exhibit surface activity and it should form self-assembled aggregates (micelles, vesicles, liquid crystalline, etc.) in liquid [22]. The most common characteristics of surfactants will be discussed below.

1.2.1. Solubilization properties

A surfactant is characterized by its tendency to adsorb at surfaces and interfaces. The term interface denotes boundary between any two immiscible phases; the term surface indicates that one of the phases is gas, usually air. Altogether five different interfaces exist:

- Solid - vapor
- Solid - liquid
- Solid - solid
- Liquid - vapor
- Liquid - liquid

The driving force for a surfactant to adsorb at an interface is to lower the free energy of that phase boundary. The interfacial free energy per unit area represents the amount of work required to expand the interface. The term interfacial tension is often used instead of interfacial free energy / unit area. Thus, the surface tension of water is equivalent to the interfacial free energy / unit area of the boundary between water and air above it. When that boundary is covered by surfactant molecules, the surface tension (or the amount of work required to expand the interface) is reduced. The denser of surfactant packing at the interface is the larger reduction in surface tension. Surfactants may adsorb at all above-mentioned types of interfaces. The tendency to accumulate at interfaces is a fundamental property of a surfactant. The degree of surfactant concentration at a boundary depends on surfactant structure and also on the nature of the two phases that meet at the interface. Therefore, there is no universally good surfactant, suiTable for all uses. The choice will depend on the application. A good surfactant should have low solubility in the bulk phases. Some surfactants (and several surface active macromolecules) are only soluble at the oil - water interface. Such compounds are difficult to handle but are very efficient in reducing of the interfacial tension. There is, of course, a limit to lower the surface and interfacial tension by the surfactant. In the normal case that limit is reached when micelles start to form in bulk solution.

Polymer solution thermodynamics has been the approach to explain the behavior of high molecular weight materials in solution. Some theories were proposed and found wide applications in polymer science and have also been quite useful in the development and understanding of polymeric solubilization [23]. The molecular structure of a surfactant is important to some surfactant characteristics as the critical micelle concentration (CMC), aggregation number, micellar shape, etc., it also controls the ability of a surfactant to solubilize a third component. The presence of solubilized additive even though the additive has no inherent surface activity can change the CMC of a surfactant substantially from that of the pure system. Hydrocarbons and polar organic compounds with low water solubility are usually found to be solubilized in the interior of the micelle or deep in the palisades layer. It was found that the amount of such materials solubilized increases as the size of the micelle increases. Considering the relative solubilizing poures of the different types of surfactant with a given hydrophobic tail, it is usually found that they can be ordered as nonionics > cationic > anionics. The supposed looser packing of the surfactant molecules in the micelles of the nonionic materials, making available more space for the incorporation of additive molecules [24]. One of the most useful properties of micellar aggregates is their ability to enhance the aqueous solubility of hydrophobic substances which otherwise are only sparingly soluble in water. The enhancement in the solubility arises from the fact that the micellar cores, for classical low mass surfactants as well as for block copolymer micelles, can serve ascompatible microenvironment for water-insoluble solute molecules. This phenomenon of enhanced solubility is referred as 'solubilization' as pointed out by Nagarajan [25].

1.2.2. Surfactants and surface tension

Surface tension is a phenomenon caused by the cohesive forces between liquid molecules. It is an effect within the surface film of a liquid that causes the film to behave like an elastic sheet [26]. Commonly it is measured in mN/m or mJ/m2 depending on how it is defined [27]. The knowledge of surface tension is useful for many applications and processes as the surface tension governs the chemical and physical behavior of liquids. It can be used to determine the quality of numerous industrial products such as paints, ink jet products, detergents, cosmetics, pharmaceuticals, lubricants, pesticides and food products. Also, it has profound effect on some steps in industrial processes such as adsorption, distillation and extraction [28, 29]. Many petroleum industry processes involve colloidal dispersions such as foams, emulsions and suspensions. All of which contain large interfacial areas, the properties of these interfaces play an important role in determining the properties of these dispersions themselves. Thus, the physical properties of the interface can be very important in all kinds of petroleum recovery and processing operations. The addition of a small quantity of a surfactant to the water would lower the surface tension and lower the amount of mechanical energy needed for foam formation. The origin of surface tension may be visualized by considering the molecules in a liquid; the attractive Van der Waals forces between molecules are felt equally by all molecules except those in the interfacial region. This imbalance pulls the latter molecules towards the interior of the liquid. The contracting force at the surface is known as the surface tension. It is known that the surface has a tendency to contract spontaneously in order to minimize the surface area. Thus, for emulsions of two immiscible liquids a similar situation applies to the droplets of one of the liquids, except that it may not be so immediately obvious which liquid will form the droplets. There will still be an imbalance of intermolecular force resulting in an interfacial tension, and the interface will adopt a configuration that minimizes the interfacial free energy.

Physically, surface tension can be considered as the sum of the contracting forces acting parallel to the surface or interface. So surface tension can be defined as the contracting force per unit length around a surface. Also there is another way to think about surface tension, the area of expansion of a surface requires energy. Since the work required expanding a surface against contracting forces is equal to the increase in surface free energy accompanying this expansion. So surface tension can be expressed as energy per unit area. There are many methods available for the measurement of surface and interfacial tension work, it should be appreciated that when solutions rather than pure liquids are involved appreciable changes can occur with time at the surfaces and interfaces, so the techniques capable of dynamic measurements tends to be the most useful [30].

When surfactant molecules adsorb at an interface they provide expanding force acting against the normal interfacial tension. Thus, surfactants tend to lower interfacial tension, this is illustrated by the general Gibbs adsorption equation for a binary isothermal system containing excess electrolyte.

$$\Gamma s = (1/RT)\,(d\gamma/\,d\ln Cs) \qquad [1.1]$$

Where Γs is the surface excess of surfactant (mol/cm^2), Cs is the solution concentration of the surfactant, and γ may be either surface or interfacial tension (mN/m). Also when surfactants concentrate in an adsorbed monolayer at a surface the interfacial film may take on any of a number of quite different properties which will provide a stabilizing influence in dispersions such as emulsions, foams and suspensions. Since polymeric surfactants consist of both hydrophobic and hydrophilic groups, they can also reduce the surface tension of aqueous phase, stabilize monomer droplets or polymer particles and can form aggregates as the result of intermolecular and/or interamolecular hydrophobic interactions [31].

1.2.3. Adsorption and confirmation of polymeric surfactants at interfaces

The process of polymer adsorption involves a number of various interactions that must be separately considered. Three main interactions must be taken into account, namely the interaction between solvent molecules with the oil in the case of O/W emulsions, which need to be displaced for the polymer segments to adsorb, the interactions between the chains and the solvent and the interaction between the polymer and the surface. Apart from knowing these interactions, one of the most fundamental considerations is the conformation of the polymer molecule at the interface. These molecules adopt various conformations, depending on their structure. For such a polymer to adsorb, the reduction in entropy of the chain as it approaches the interface must be compensated by energy of adsorption between the segments and the surface. In other words, the chain segments must have minimum adsorption energy, otherwise no adsorption occurs. With polymers that are highly water soluble, such as poly (ethylene oxide) (PEO), the interaction energy with the surface may be too small for adsorption to occur, and if this takes place the whole molecule may not be strongly adsorbed to the surface. So many commercially available polymers that are described as homopolymers, such as poly (vinyl alcohol) (PVA) contain some hydrophobic groups or short blocks (vinyl acetate in the case of PVA) that ensured their adsorption to hydrophobic surfaces [32].

Several theories exist that describe the process of polymer adsorption, which have been developed either using statistical mechanical approach or quasi-lattice models. In the statistical mechanical approach, the polymer is considered to consist of three types of structures with different energy states, trains, loops and tails. The structures close to the surface (trains) are adsorbed with an internal partition function determined by short-range forces between the segment and surface [33].

1.3. Micillization Process

At low concentrations, most properties of a surfactant are similar to those of a simple electrolyte. One noTable exception is the surface tension, which decreases rapidly with surfactant concentration. At a higher concentration, which is varied for different surfactants, unusual changes are recorded. For example, the surface tension takes on an approximately constant value, while light scattering starts to increase and self-diffusion starts to decrease. All these observations are consistent with a change-over from a solution containing single surfactant molecules or ions, unimers, to a situation where the surfactant occurs more and more in a self-assembled or self associated state. The first-formed aggregates are generally approximately spherical in shape. The concentration where the aggregates micelles start to form is called the critical micelle concentration, abbreviated as CMC. The CMC is the most important characteristic of a surfactant in consideration of the practical uses. The two most common and generally applicable techniques, for measuring the CMC are surface tension and solubilization, the solubility of an otherwise insoluble compound. For an ionic amphiphile, the conductivity offers a convenient approach to obtain the CMC. However, as a very large number of physicochemical properties are sensitive to surfactant micellization, there are numerous other possibilities, such as self-diffusion measurements and NMR (nuclear magnetic resonance) and fluorescence spectroscopy. The CMC is not an exactly defined quantity, which causes difficulties in its determination. For long-chain amphiphiles, an accurate determination is straightforward and different techniques give the same results. However, great care must be taken not only in the measurements but also in evaluating the CMC from the experimental data.

The surfactants concentration in solution affects their behavior, thus in aqueous solutions dilute concentrations of surfactant act much as normal electrolytes, but at higher concentrations it shows a different behavior. This behavior is explained in terms of the formation of organized aggregates of large numbers of molecules called micelles, in which lipophilic parts of the surfactants associate in the interior of the leaving hydrophilic parts to face the aqueous medium. The formation of micelles in aqueous solutions is generally viewed as a compromise between the tendency for alkyl chains to avoid energetically unfavorable contacts with water and the desire for the polar parts to maintain contact with the aqueous environment. Due to the presence of the hydrophobic effect, surfactant molecules adsorb at interfaces, even at low surfactant concentrations. As there will be a balance between adsorption and desorption (due to thermal motions), the interfacial condition requires some time to establish. The surface activity of surfactants should therefore be considered a dynamic phenomenon. This can be determined by measuring surface or interfacial tensions versus time for a freshly formed surface [30].

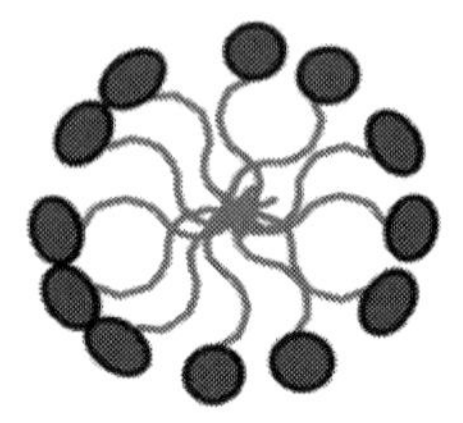

A micelle is the lipophilic ends of the surfactant molecules dissolve in the oil, while the hydrophilic charged ends remain outside and shielding the rest of the hydrophobic micelle.

Surfactants reduce the surface tension of water by adsorbing at the liquid-gas interface. They also reduce the interfacial tension between oil and water by adsorbing at the liquid-liquid interface. Many surfactants can also assemble in the bulk solution into aggregates. Some of these aggregates are known as micelles. The concentration at which surfactants begin to form micelles is known as the critical micelle concentration or CMC. When micelles form in water, their tails form a core that is like an oil droplet, and their (ionic / polar) heads form an outer shell that maintains favorable contact with water. When surfactants assemble in oil, the aggregate is referred to as a reverse micelle. In a reverse micelle, the heads are in the core and the tails maintain favorable contact with oil [34].

1.3.1. Critical micelle concentration

The CMC is a property of the surfactant since micellization is opposed by thermal and electrostatic forces. A low CMC is favored by increasing the molecular mass of the lipophilic part of the molecule, lowering the temperature and adding electrolyte. The physico-chemical properties of surfactants vary above and below a specific surfactant concentration. The CMC value is defined in terms of the pseudo-phase model as the concentration of the maximum solubility of the monomers in that particular solvent. The concept of the CMC value is very useful when discussing the association of surfactants into micelles. So the CMC value is perhaps the most frequently measured and discussed micellar parameter. In the most petroleum industry surfactant applications such as improving oil recovery processes, surfactant must be usually present at a concentration higher than the CMC because the greatest effect of the surfactant whether in interfacial tension lowering or in promoting foam stability is achieved when a significant concentration of the micelle is present. The CMC is also of interest because at concentration above this value the adsorption of surfactants onto reservoir rock surfaces increases very little. So the CMC represents the solution concentration of surfactant from which nearly maximum adsorption occurs [24, 35].

1.3.2. CMC Measurement

There are some common methods for measuring CMC value of a surfactant micelle. Some of these methods are collected as:

- UV/Vis IR spectroscopy
- Fluorescence spectroscopy
- Nuclear magnetic resonance spectroscopy
- Electrodepotential conductivity
- Surface tension
- Foaming

Most methods have been developed for a relatively small set of pure surfactants involving very dilute electrolyte solutions and only ambient temperature and pressure. The determination of CMC at elevated temperature and pressure is much more difficult than for ambient conditions. Most high temperature CMC studies have been by conductivity measurements and have therefore limited to ionic surfactants. For petroleum industry processes one tends to have a special interest in the CMC of partical surfactants that may be anionic, cationic and non-ionic or amphteric [30].

1.3.3. Factors affecting micellization process

There are many factors affecting CMC, the most important will be discussed below.

1.3.3.1. Effect of hydrophilic and hydrophobic groups

The length of the chain of a hydrocarbon surfactant has been shown to be a major factor determining the CMC. It is known that the CMC decreases logarithmically as the number of carbons in the chain of a homologous series n_c increases Equation [1.2]. For straight chain hydrocarbon surfactants of about 16 carbon atoms or less bound to a single terminal head group, the CMC is usually reduced to approximately one half of its previous value with the addition of each methylene group. For non-ionic surfactants, the effect can be much larger, with a decrease by a factor of 10 following the addition of two carbons to the chain [36]. The insertion of a phenyl and other linking groups, branching of the alkyl group, and the presence of polar substituent groups on the chain can produce different effects on the CMC.

$$\text{Log}_{10}\,\text{CMC} = \text{A-Bn}_c \qquad [1.2]$$

The nature of the hydrophobic group has a major effect on the CMC of a surfactant. The effect of the hydrophilic head group on the CMC of a series of surfactants with the same hydrocarbon chain may also vary considerably, depending on the nature of the change. Thus in aqueous solution the difference in CMC for C_{12} hydrocarbon with an ionic head group will lie in the range of 0.001 M while a non-ionic material with the same chain will have a CMC in the range of 0.0001M, Table (1.2) shows the CMC values of several ionic surfactants. The nature of the ionic head group has a rather small effect such a result is not surprising in view of the fact that the primary driving force in favour of micelle formation is the energy gain due to reduction of water-hydrophobic interactions. While the effect of the ionic group, beyond its impact on water solubility, is to work against aggregation process. Also the location of the head group along the hydrophobic chain can greatly affect micellization. It has been shown, for example, that as the charge on the hydrophilic groumoved away from α – carbon of the hydrophobe, the CMC will decrease [37].

1.3.3.2. Effect of pH and temperature

Since most modern, industrial important surfactants consist of long alkyl chain salts of strong acids, it might be expected that solution pH would have a relatively small effect. In solutions of sulfonate and sulfate salts, where the concentration of acid or base significantly exceeds that of the surfactant, the excess will act as if it were simply neutral electrolyte with roughly the same results. It is expected that pH will have no effect on the CMC of non-ionic surfactants, but at very low pH it is possible that the protonation of the ether oxygen of EO surfactants could occur [38].

The effects of temperature changes on the CMC of surfactants in aqueous solution have been found to be quite complex. It has been shown for example, that the CMC of most ionic surfactants passes through a minimum as the temperature is varied from about 0 through 333-343 K. Non-ionic and zwitterionic materials are not quite so predicTable, although it has been found that some nonionics reach a cmc minimum around 323 K. The temperature dependence of the CMC of polyoxyethylene nonionic surfactants is especially important since the head group interactions essentially totally hydrogen bonding in nature. The solubilization of such materials in aqueous solution is commonly found to exhibit an inverse temperature / solubility relationship. A major manifestation of such a relationship is the presence of the so-called cloud point for many non-ionic surfactants [39].

Table (1.2): The Effect of the Hydrophilic Group on the CMC of Surfactants with Common Hydrophobes

Hdrophobe	Hydrophile	Temperature ^{O}C	CMC(mM)
$C_{12}H_{25}$	COOK	25	12.5
$C_{12}H_{25}$	SO_3Na	25	8.1
$C_{12}H_{25}$	NH_3Cl	30	14
$C_{12}H_{25}$	$N(CH_3)_3Cl$	30	20
$C_{16}H_{23}$	NH_3Cl	55	0.85

$C_{16}H_{23}$	$N(CH_3)_3Cl$	30	1.3
$C_{16}H_{23}$	$N(CH_3)_2C_2H_4OHCl$	30	1.2
$C_{16}H_{23}$	$N(CH_3)(C_2H_4OH)_2Cl$	30	1.0
C_8H_{17}	OCH_2CH_2OH	25	4.9
C_8H_{17}	$(OCH_2CH_2)_2OH$	25	5.8
C_9H_{19}	$COO(CH_2CH_2O)_9CH_3$	27	1.0
$C_{10}H_{21}$	$O(CH_2CH_2O)_8CH_3$	30	0.6

1.3.4. Hydrophile - lipophile balance (HLB) concept

An emulsion is a dispersion of two immiscible liquids, normally referred to as oil and water. Under intense agitation fine droplets of one phase in the other may be formed. Such dispersants are not stable without any stabilizing agent. The droplets will coalesce and the coalescing process normally starts immediately as agitation is stopped. In order to enhance the dispersion process and to stabilize the emulsion, emulsifying agents and/or emulsion stabilizers are added. Water-soluble polymers are often used as stabilizers of emulsions formed by the action of low molecular weight surfactants, i.e. emulsifiers. They offer further stabilization through steric forces between adsorbed polymer layers on the dispersed droplets and are referred to protective colloids. The finely divided hydrophobic particles accumulate at the oil-water interface. They act through their wetting properties and it has been seen that the contact angle that the meniscus makes with the particle is critical. Only a relatively small change in contact angle may change the fine particle from being an emulsion stabilizer to an emulsion breaker. The presence of polymers or particles or a combination of the two as emulsion stabilizers is very common in biological emulsions [40]. The rule of thumb in emulsion technology is that water-soluble emulsifiers stabilize O/W emulsions meanwhile oil-soluble emulsifiers stabilize W/O emulsions. This concept is known as Bancroft's rule and dates back to the beginning of this century. Bancroft's rule is entirely qualitative. In an attempt to extend it into some kind of quantitative relationship between surfactant hydrophilicity and function in solution, Griffin in 1949 introduced the concept of hydrophil-lipophil balance (HLB) of a surfactant. HLB numbers for normal non-ionic surfactants were determined by simple calculation [41]:

1- For alcohol ethoxylates and alkylphenol ethoxylates:

$$HLB = \frac{weight\ \%\ \text{ethylene oxide}}{5} \qquad [1.3]$$

2- For polyol ethoxylate:

$$HLB = \frac{weight\ \%\ \text{ethylene oxide} + \text{weight}\ \%\ \text{polyol}}{5} \qquad [1.4]$$

3- For fatty acid esters of polyol:

$$HLB = 20\left[1 - \frac{saponification\ \text{number}}{\text{acid number}}\right] \qquad [1.5]$$

Davies, who introduced a Scheme to assign HLB group numbers to chemical groups composing a surfactant, later extended Griffin's HLB number concept. Davies' formula and typical group numbers are shown in Table (1.3). From the Table it can be seen, for instance, that sulfate is much more potent polar group than carboxylate and that a terminal hydroxyl group in poly (oxyethylene) chain is more powerful in terms of hydrophilicity than a sugar hydroxyl group. The HLB number concept, and particularly Griffin's version (which is restricted to non-ionics), has proved useful as a first selection of surfactant for a given application. Table (1.3) shows how the appearance of aqueous surfactant solutions depends on surfactant HLB. It also indicates typical applications of surfactants of different HLB number intervals. It can be seen that an emulsifier for a w/o emulsion should be hydrophobic with an HLB number of 3-6 and an emulsifier for an o/w emulsion should be in the HLB number range of 8-18. This is obviously in line with Bancroft's rule. For room temperature operations, the HLB numbers calculated according to Griffin (or Davies) give a useful prediction for emulsifier selection. However, problems arise if the temperature is raised during emulsification or when the ready-made emulsion is stored at very low temperatures.

Non-ionic surfactants of polyoxyethylene type are very temperature sensitive. Many of them give O/W emulsions at ambient conditions and W/O emulsions at elevated temperature. Also, factors such as electrolyte concentration in the water, oil polarity and the water-to-oil ratio may also influence what type of emulsion forms. Evidently, the HLB number cannot be used as a universal tool to select the appropriate emulsifier or to determine what type of emulsion will form with a specific surfactant.

Table (1.3): Determination of HLB number according to Davies

1. Hydrophilic group numbers	
-SO_4Na	35.7
-CO_2K	21.1
-CO_2Na	19.1
-N (tertiary amine)	9.4
Ester (sorbitan ring)	6.3
Ester (free)	2.4
-CO_2H	2.1
-OH (free)	1.9
-O-	1.3
-OH (sorbitan ring)	0.5
2. Lipophilic group numbers	
-CF_3	-0.870
-CF_2-	-0.870
-CH_3	-0.475
-CH_2-	-0.475
-CH-	-0.475

HLB=7+Σ(hydrophilic group numbers)+Σ(lipophilic group numbers)

1.4. Polymeric Surfactants Based on Polyoxyalkaylene

A diverse set of compounds containing repeating units of ethylene oxide (EO), propylene oxide (PO) and butylene oxide (BO) have been synthesized on a commercial scale. By varying the number of monomer units present and their sequence in the macromolecule, polymers of different properties have been prepared. Of chief interest are the A-B-A type block copolymers where A = polyethylene oxide (PEO) and B = polypropylene oxide (PPO), commercially known as Pluronics® or Poloxamers. The pendant methyl groups ($-CH_3$) of the PPO units make the central core hydrophobic and the PEO chains demonstrate hydrophilic behavior. Due to this amphiphilic nature, all of the Pluronics® demonstrate surface activity. This property has been exploited to their classification and commercial use as nonionic surfactants, and depending upon the ratio of EO/PO groups present (also known as the hydrophile-lipophile balance) these molecules are used as detergents, foaming agents, foam suppressors, emulsifiers and wetting agents. In aqueous solutions PEG (PEO) binds to 2-3 water molecules per ethylene oxide unit [42]. Additionally, entropic effects lead to aggregation of the central hydrophobic PPO of the triblocks to minimize their exposure to the surrounding water molecules. This leads to formation of micelles, vesicles and other self-assembled structures in solution. When exposed to a hydrophobic surface, these molecules adsorb on such a surface via their hydrophobic core, forming monolayers [43]. Depending upon the surface density, these layers assume a pancake, mushroom or brush-type configuration, and make the underlying surface hydrophilic. This property of self-assembly coupled with their non-toxicity has found several uses for Pluronics® in the medical field.

Polyoxyethylene (PEO)–polyoxypropylene (PPO) block copolymers are widely used in practice. They have several advantages: non-toxicity; degradability; good stabilization ability; and their production are easy and relatively inexpensive. Because of their multiple site attachment at surfaces, the adsorbed polymers are less accessible to desorption or displacement compared to low molecular surfactants. The block copolymers are used as stabilizers of dispersions, foams and emulsions in the pharmaceutical and cosmetic industries [44].

Commercial demulsifiers are polymeric surfactants such as copolymers of polyoxyethylene and polypropylene or polyester or blends of different surface-active substances [45]. The first anionic surfactants used as demulsifiers are known as soaps and are usually prepared by saponification of natural fatty acid glycerides in alkaline solution. The degree of water solubility is controlled by the length of the alkyl chain ranging from 12 to 18. Amphiphilic block copolymers with dendrimer structures; dendrimers are defined as macromolecules containing highly branched and dimensional structures with a large number of reactive terminal groups [46]. Water-soluble amphiphilic dendrimers have attracted considerable attention due to their unique properties and as new polymeric materials for applications in many areas such as preparation of micelles, liquid crystals, and molecular encapsulation [47]. Amphiphilic dendrimer copolymers can be used as demulsifiers. In amphiphilic polymers, (EO) increases hydrophilicity whereas (PO) increases hydrophobicity. A combination of the two groups with a suiTable degree of polymerization will give intermediate solubility in water and oleic phases. Because the dendimers have good penetrability, they can quickly enter bulk solution and then pervade the oil/water interface. The interfacial activity of the dendrimers is higher than that of natural surfactants. So they can easily displace the natural surfactants, with hydrophilic PEO chains penetrating into water drops while hydrophobic PPO chains remain in the oleic phase, so that, these dendritic macromolecules exhibited excellent demulsification performance for crude oil emulsion.

The simplest type of a polymeric surfactant is a homopolymer that is formed from the same repeating units: (PEO) and poly (vinylpyrrolidone) (PVP). Homopolymers have little surface activity at oil / water (O/W) interfaces. Polymeric surfactants of the block (A-B or A-B-A) are also known. A block copolymer is a linear arrangement of blocks of varying compositions [23]. Generally, block copolymers are defined as macromolecules with linear and/or radial arrangement of two or more different blocks of varying monomer composition [48]. Block copolymers exhibit surface activity since one of the block is soluble in one of the phases and the other immiscible in the other phase (e.g., A-B block, A hydrophilic and B hydrophobic).

To study more extensively the influence of intramolecular interaction on the solutions properties, it has been reported that the hydrophobic interaction of a water-soluble amphiphilic polymer is influenced not only by the chemical structure and composition of the hydrophobic groups but also by the distribution of these groups along the chain [49]. Block and graft copolymers having the general formulae, L - $(L)_m$ - L - H - $(H)_n$ – H Where L and H are the lipophilic and the hydrophilic moieties respectively. In random copolymer, L and H may not be segregated sufficiently (in analogy to monomeric surfactants) and hence are not expected to exhibit surface activity [50]. Indeed, it was reported that when propylene oxide (L) and ethylene oxide (H) were allowed to copolymerize randomly, the products did not possess the essential characteristics of surfactants [51].

Preparation of polymeric surfactants and their application have been reported in details [52].The emergence of synthetic, as opposed to natural, polymeric surfactants [53] is a result of the increasing technical development in both polymers and surfactants [54]. The preparation and properties of amphiphilic graft copolymers containing poly (ethylene oxide) grafts has been the subject of recent investigations [55]. The two major varieties of non-ionic polymeric surfactants are the poly (alkylene oxide) block copolymers and the phenolic polymers. Block copolymers containing poly (propylene oxide) as a hydrophobic block and poly (ethylene oxide) as a hydrophilic block were synthesized and tested as non-ionic surfactants [56]. Their versatility arises from the flexibility, and their properties can be modified through variations in: the total molecular weight, the ratio of molecular weights of the blocks, the arrangement of the blocks and the type of initiator used.

The synthesis, properties and applications of block copolymers have been reviewed in detail [57]. The relation between surface activity and molecular structure of nonionic polymeric surfactants synthesized by grafting of polypropylene glycol onto hydrophilic polymer has been investigated by Nakamura [58].The preparation and properties of comb-shaped amphiphilic copolymers composed of hydrophobic polymethacrylate backbone and hydrophilic poly (ethylene glycol) side chains were reported [59].

Amphiphilic block copolymers are mainly di- or tri-block copolymers where the different blocks are incompatible, therefore giving the polymer its unique properties. The most extensively studied and industrially significant amphiphilic polymers contain PEG or PEO as hydrophilic segment. In terms of hydrophobic segments, the most generally used polymers are poly (propylene oxide), it was reported that the first amphiphilic block copolymer were prepared in the early 1950s by Lundsted on the basis of ethylene oxide and propylene oxide [60]. It was found that wettability increased with an increase in the length of the PEO block, while the best surface activity was achieved with the (PPO) block. Pluronics were shown to exhibit emulsifying properties, phase transfer catalyst properties, complixing abilities with alkali salts and also ionic conducting behavior. Recently it was also found that Pluronics prevented protein adsorption and platelet adhesion [61]. Since the introduction of Pluronics into the market, various advances have been made in the synthesis of amphiphilic copolymers. The advances have been reviewed [60].

Amphiphilic block copolymers with dendrimer structures were synthesized by anion polymerization and these dendritic macromolecules exhibited excellent demulsification performance for crude oil emulsion. According to the demulsification experiments, it was determined that the copolymers became more efficient with more complex molecular structure due to increased penetrability. The physical model of demulsification intelligibly explained the micromechanism of flocculation and coalescence of water drops in emulsion because of the good adsorption and displacement behaviors of dendrimers [62].

The use of amphiphilic copolymers in separation systems is attracting increasing interest. One reason for this is the recent advances in the physicalchemical characterisation of the solution behaviour of amphiphilic copolymers. The knowledge of how these polymers behave opens the way to applications. For block copolymers of the type PEO-PPO-PEO the interest has been focused mainly on the extraction of small hydrophobic solutes. The micelle formation exhibited by these polymers offers the possibility to solubilize hydrophobic compounds in the hydrophobic interior of the micelles. This means that amphiphilic copolymers in aqueous solutions are potential alternatives to organic solvents for extractions of organic molecules. Water solutions of copolymers are environmentally friendly solvents, and it will describe research on extractions using amphiphilic copolymer systems [63].

Demulsifiers provide an important means of breaking water-in-oil emulsions that occur in industrial processes. The properties and the performance of 20 blocked copolymers from four surfactant families were investigated and three pairs isomeric compounds were compared. The results show that different positions of the (EO) and (PO) in block copolymers lead to different hydrophile–lipophile balances (HLB) of surfactant. The sequential block copolymer is more hydrophilic than the reverse-sequential one with similar chemical composition. Generally, the demulsification performance of sequential copolymers is better than that of reverse-sequential copolymers. Position isomerism of the surfactant affects demulsification performance by changing the hydrophile–lipophile balance, interfacial properties, and steric characteristics at the interface [64]. Four poly(ethylene oxide)–block–poly(propylene oxide)–block–poly(ethylene oxide) copolymers with different molecular weights and PPO/PEO composition ratios were synthesized. The characterization of the PEO–PPO–PEO triblock copolymers was studied by surface tension measurement, UV–vis spectra, and surface pressure method.

The low solubility in biological fluids displayed by about 50% of the drugs still remains the main limitation in oral, parenteral, and transdermal administration. Among the existing strategies to overcome these drawbacks, inclusion of hydrophobic drugs into polymeric micelles is one of the most attractive alternatives. Amphiphilic poly(ethylene oxide)–poly(propylene oxide) block copolymers are thermoresponsive materials that display unique aggregation properties in aqueous medium. Due to their ability to form stable micellar systems in water, these materials are broadly studied as hydrosolubilizers for poorly water-soluble drugs. Block copolymers with segments of either poly(ethylene oxide), poly(propylene oxide), or mixtures of poly(ethylene oxide)/poly(propylene oxide) and monodisperse aramide segments were discussed [65].

The length of the polyether segments as well as the concentration of polyethylene oxide was varied. The crystallinity of the monodisperse aramide segments was found to be high and the crystals, dispersed in the polyether phase, displayed nano-ribbon morphology. The PEO segments were able to crystallize and this crystalline phase reduced the low-temperature flexibility. The PEO crystallinity and melting temperature could be strongly reduced by copolymerization with PPO segments. By using mixtures of PEO and PPO segments, hydrophilic copolymers with decent low-temperature properties could be obtained [66]. Block copolymers of (EO) and (PO) are characterized by combination of two-dimensional chromatography and matrix-assisted laser desorption/ionization time-of-flight mass spectrometry (MALDI-TOF-MS). Liquid chromatography under critical conditions (LCCC) is used as first dimension and fractions are collected, mobile phase evaporated and diluted in the mobile phase used in second dimension (SEC or LAC). This two-dimensional chromatography in combination of MALDI-TOF-MS gives information about purity of reaction products, presence of the byproducts, chemical composition and molar mass distribution of all the products [67].

MALDI-TOF-MS was used to analyse the block length of commercially available block copolymers of poly(ethylene oxide) and poly(propylene oxide) (PEO-*b*-PPO) based on the fragmentation behaviour in collision induced dissociation (CID) experiments. Therefore, a step-by-step procedure was used starting with PEG and PPG standards, PEG-PPG blends and endgroup-functionalized PPGs, to understand the fragmentation behaviour of the different species. These results showed that characteristic fragment patterns of the homopolymers and PEG-PPG mixtures can be obtained that facilitate the interpretation of the fragment spectra of PEO-*b*-PPO di- and triblock copolymers. It was found that di- and triblock copolymers can be differentiated by their fragment spectra. In addition, the sequence of monomer units in the diblock copolymers could be determined [68].

The association properties (PEO-PPO-PEO) copolymers (commercially available as Poloxamers and Pluronics) in aqueous solutions and the adsorption of these copolymers at interfaces are reviewed [69]. At low temperatures and/or concentrations the PEO-PPO-PEO copolymers exist in solution as individual coils (unimers). Thermodynamically stable micelles are formed with increasing copolymer concentration and/or solution temperature, as revealed by surface tension, light scattering, and dye solubilization experiments. The unimer-to-micelle transition is not sharp, but spans a concentration decade or 10 K. The critical micellization concentration (CMC) and critical micellization temperature (CMT) decrease with an increase in the copolymer PPO content or molecular weight. The dependence of CMC on temperature, together with differential scanning calorimetry experiments, indicated that the micellization process of PEO-PPO-PEO copolymers in water is endothermic and driven by a decrease in the polarity of (EO) and (PO) segments as the temperature increases, and by the entropy gain in water when unimers aggregate to form micelles (hydrophobic effect). The free energy and enthalpy of micellization can be correlated to the total number of EO and PO segments in the copolymer and its molecular weight. The micelles have hydrodynamic radii of approximately 10 nm and aggregation numbers in the order of 50. The aggregation number is thought to be independent of the copolymer concentration and to increase with temperature. In addition, the lattice models can provide information on the distribution of the EO and PO segments in the micelle. The PEO-PPO-PEO copolymers adsorb on both air-water and solid-water interfaces; the PPO block is located at the interface while the PEO block extends into the solution, when copolymers are adsorbed at hydrophobic interfaces. Gels are formed by certain PEO-PPO-PEO block copolymers at high concentrations, with the micelles remaining apparently intact in the form of a "crystal". The gelation onset temperature and the thermal stability range of the gel increase with increasing PEO block length [69].

Block copolymers consisting of (PEO) / (PPO) can self-assemble in water and water/oil mixtures (where water is a selective solvent for PEO and oil a selective solvent for PPO) to form thermodynamically stable spherical micelles as well as an array of lyotropic liquid crystalline mesophases of varying morphology. Significant advances have been made over the past year on the identification of different morphologies, the delineation of the composition-temperature ranges where they occur, and the structural characterization of the morphologies using primarily small angle scattering techniques. Important new findings on the copolymer micellization in water as affected by cosolutes, and on the time-dependency of the surface activity have also been reported [70].

Four poly(ethylene oxide)–block–poly(propylene oxide)–block–poly(ethylene oxide) copolymers with different molecular weights and PPO/PEO composition ratios were synthesized. The characterization of the PEO–PPO–PEO triblock copolymers was studied by surface tension measurement, UV–vis spectra, and surface pressure method. These results clearly showed that the CMC of PEO–PPO–PEO was not a certain value but a concentration range, in contrast to classical surfactant, and two breaks around CMC were reflected in both surface tension isotherm curves and UV–vis absorption spectra. The range of CMC became wider with increasing PPO/PEO composition ratio. Surface pressure π–A curves revealed that the amphiphilic triblock copolymer PEO–PPO–PEO molecule was flexible at the air/water interface. We found that the minimum area per molecule at the air/water interface increased with the proportion of PEO chains. The copolymers with the same mass fractions of PEO had similar slopes in the isotherm of the π–A curve. From the demulsification experiments a conclusion had been drawn that the dehydration speed increased with decreased content of PEO, but the final dehydration rate of four demulsifiers was approximate. We determined that the coalescence of water drops resulted in the breaking of crude oil emulsions from the micrograph [71].

Aiming at developing new reverse thermo-responsive polymers, poly(ethylene oxide)-poly(propylene oxide) multiblock copolymers were synthesized by covalently binding the two components using carbonyl chloride and diacyl chlorides as the coupling molecules. The appropriate selection of the various components allowed the generation of systems displaying much enhanced rheological properties. For example, 15 wt% aqueous solutions of an alternating poly(ether-carbonate) comprising PEO6000 and PPO3000 segments, achieved a viscosity of 140,000 Pa s, while the commercially available Pluronic F127 displayed 5,000 Pa s only. Furthermore, the structure of the chain extender played a key role in determining the sol–gel transition. While poly(ether-ester)s containing therephtaloyl (*para*) and isophtaloyl (metha) coupling units failed to gel at any concentration, a 15 wt% aqueous solution of the polymer chain-extended with phtaloyl chloride (*ortho*) gelled at 316 K. The water solutions were also studied by dynamic light scattering and a clear influence of the PEO/PPO ratio on the aggregate size was observed. By incorporating short aliphatic oligoesters into the backbone, prior to the chain extension stage, reverse thermal gelation-displaying biodegradable poly(ether-ester-carbonate)s, were generated [72].

The association behavior of (PEO–PPO–PEO) block copolymers in aqueous solution with hexadecyltrimethyl ammonium bromide (HTAB), tetradecyltrm-ethylammonium bromide (TTAB) and dimethylene bis(decyldimethylamm- onium bromide) (10–2–10), was studied by fluorescence, viscosity, and Krafft temperature measurements. It has been observed that $(EO)_{18}(PO)_{31}(EO)_{18}$ interacts more strongly than $(EO)_2(PO)_{15.5}(EO)_2$ and $(EO)_{2.5}(PO)_{31}(EO)_{2.5}$ with HTAB/TTAB due to synergistic interactions. A stronger capability of $(EO)_{18}(PO)_{31}(EO)_{18}$ to interact with cationic surfactants arises from the greater number of electronegative EO units (total 36 EO units) than of $(EO)_2(PO)_{15.5}(EO)_2$ (total 4 EO units) and $(EO)_{2.5}(PO)_{31}(EO)_{2.5}$ (total 5 EO units). The antagonistic mixing behavior of present triblock polymers has been observed with 10–2–10. A difference in the mixing behavior of the latter from that of HTAB/TTAB has been attributed to its dimeric nature, which may create steric hindrances with triblock polymer components at the head group region in the mixed state [73]. Adsorption of three commerciably available ethylene oxide (EO)/propylene oxide (PO) block copolymers with the general formula $EO_aPO_bEO_a$, was studied with ellipsometry on a hydrophobicity gradient surface with contact angles ranging from 0 to 90°. Generally, higher adsorption values were obtained on the hydrophobic side than on the hydrophilic. The isotherm shapes were, strictly speaking non-Langmuir. However, within the range 0–10 ppm the isotherms could reasonably well be fitted to a Langmuir model. Generally, semi-plateau values of around 1.5–2.0 mg m^{-2} were measured on the hydrophobic side and around 0.0–0.5 mg m^{-2} on the hydrophilic, and a slowly progressing increase of the adsorption was observed at higher concentrations. The most hydrophilic surfactant, $EO_{148}PO_{56}EO_{148}$, gave slightly lower values on the hydrophobic side compared to the two others ($EO_{98}PO_{69}EO_{98}$ and $EO_{37}PO_{56}EO_{37}$). In contrast, comparatively higher affinity and adsorption capacity was found for $EO_{148}PO_{56}EO_{148}$ on the hydrophilic side of the gradient possibly indicating weak attraction between EO and surface SiOH groups. The adsorption capacity increased only slightly with temperature, but the adsorption affinity increased significantly between 20 and 38°C for $EO_{98}PO_{69}EO_{98}$, especially on the hydrophilic side. No spontaneous desorption was observed for any of the EO/PO copolymers. On the other hand, when exposed to an albumin solution, significant desorption and displacement occurred, in particular on the the hydrophobic end of the gradient as verified by ESCA analysis [74]. A cellulose membrane and nude mouse skin with series concentrations of PEO–PPO–PEO block copolymers were used to examine the sustained-release pattern and permeation of fentanyl. The in vivo percutaneous absorption was examined using rabbits to evaluate the preliminary pharmacokinetics of fentanyl with 46% PEO–PPO–PEO copolymer formulation patches. The micelle formation ability of this block copolymer and the penetration ability of PEO–PPO–PEO copolymer over time were also studied by pyrene fluorescence probe methods and the dynamic light scattering test. At a concentration of 46% at 37°C, PEO–PPO–PEO copolymers formed a gel and showed a pseudo-zero-order sustained-release profile. With increasing concentration of copolymer in the cellulose membrane transport, the apparent release flux of fentanyl (200 μg/ml) decreased to 1.09±0.19 μg cm^{-2} h^{-1}. Assessment of the effect of the copolymer on nude mouse skin also showed a decrease in the apparent permeability coefficient [(P_{H2O})=2.24±0.47×10^{-6} cm s^{-1} vs. (P_{46}% block $_{copolymer}$)=0.93±0.23×10^{-7} cm s^{-1}]. The preliminary pharmacokinetics of the fentanyl patch was shown to be in steady state within 24 h, and this was maintained for at least 72 h with an elimination half-life ($t_{1/2}$) of 10.5±3.4 h. A fluorescence experiment showed polymeric micelle formation of PEO–PPO–PEO copolymers at 0.1% (w/w) within 50 nm micelle size and the PEO–PPO–PEO copolymers were able to penetrate nude mouse skin within 24 h. Thus, it appears that fentanyl preparations based on PEO–PPO–PEO copolymer gel might be practical for percutaneous delivery [75].

The specific behavior of PEO–PPO–PEO tri-block copolymers in aqueous solutions was studied in the presence of vinyl acetate (VAc) and colloidal silica. Several factors controlling specific interactions were investigated, the block copolymer molecular weight and hydrophilic/hydrophobic balance (changed via the PEO/PPO ratio), temperature and silica content. Emulsion formation for water/VAc/copolymer mixtures was investigated for different water/VAc ratios, three temperatures (293, 313 and 333 K) and three block copolymer concentrations. It was proved that silica addition in the VAc/water/PEO–PPO–PEO tri-block copolymer mixtures results in a synergetic effect of the (copolymer-silica) system that act as a stabilizer for the emulsion droplets. Depending on the block copolymer content and HLB as well as on the temperature, emulsion droplets with different diameters were obtained. Efficient encapsulation of VAc can be promoted using silica particles and PEO–PPO–PEO block copolymers with a proper choice of the polymer and temperature [76]. Recently, interest has been focused on studying their roles in reducing nonspecific protein adsorption and cell adhesion on biomaterial/biosensor surfaces. Although the ability of the adsorbed PEO–PPO–PEO triblock copolymer to reduce protein adsorption has been observed frequently, its detailed mechanism of functioning is yet to be clarified. In order to delineate this detailed mechanism, one first needs to know the adsorption behavior of Pluronics on various substrates. Although gold is a commonly used substrate for the probes of various biosensors, there is no direct data reporting the adsorption behavior of Pluronics on it. In this study, we use surface plasma resonance (SPR) technique and Ellipsometry to detect the adsorption isotherms of various Pluronics on a gold surface. The adsorption isotherm of Pluronics of P103, P104 and F108 all exhibit two plateaus that corresponding to desorption of the EO and the micellization in bulk solution. At bulk concentration of 1×10^{-2} mg/ml, the total adsorbed amount of the copolymer increases with the increase of the molar mass of the polymer and with the decrease of the EO/PO ratio that suggests the strong influence of the hydrophobic PPO on the adsorbed amount of the polymer. At the CMC, the plateau value of adsorbed amount of Pluronics on Au is independent of molar mass and EO/PO ratio of copolymer. Protein adsorption on gold surfaces modified with various Pluronics shows good correlation between the amounts of protein adsorbed and the value of the reduced surface coverage, σ^* of the Pluronic adlayer. In general, by increasing σ^*, the nonspecific adsorption of human serum albumin can be reduced [77].

Star-shaped copolymers based on the linear core poly(ethylene oxide)-poly(propylene oxide) triblock (PEO-b-PPO-b-PEO) and poly(methyl methacrylate) (PMMA) arms were synthesized *via* a divergent approach. The PEO based chain was firstly dendronized with a polyester system that increased the number of hydroxyl groups per each terminal chain affording cores of generation 1 (G1, two hydroxyl groups *per* terminal chain) and 2 (G2, four hydroxyl groups *per* terminal chain). These were used as macroinitiators in the Atom Transfer Radical Polymerization (ATRP) of methyl methacrylate. The copolymers, characterized by different number and length of PMMA blocks, were studied by ^{1}H-NMR, GPC and DSC. A study of influence of architecture on physical properties such as mechanical properties, water sorption and oxygen permeability, was also performed. It was found that both oxygen permeability and water sorption are enhanced with respect to the PMMA homopolymer by a magnitude influenced by the structure. Differently, the values of Young's modulus remained comparable to that of PMMA [78].

The synthesis and characterisation of segmented block copolymers based on mixtures of hydrophilic poly(ethylene oxide) and hydrophobic poly(tetramethylene oxide) polyether segments and monodisperse crystallisable bisester tetra-amide segments are reported. The PEO length was varied from 600 to 8000 g/mol and the PTMO length was varied from 650 to 2900 g/mol. The influence of the polyether phase composition on the thermal mechanical and the elastic properties of the resulting copolymers were studied. The use of high melting monodisperse tetra-amide segments resulted in a fast and almost complete crystallisation of the rigid segment. The copolymers had only one polyether glass transition temperature, which suggests that the amorphous polyether segments were homogenously mixed. Thermal analysis of the copolymers showed one polyether melting temperature that was lower than in the case of ideal co-crystallisation between the two polyether segments. However, at PEO or PTMO lengths larger than 2000 g/mol two polyether melting temperatures were observed. The copolymer with the best low temperature properties was based on a mixture of PEO and PTMO segments, both having a molecular weight of 1000 g/mol, at a weight ratio of 30/70 [79].

1.5. Applications of Surfactants in Petrolium Field

Surfactants are widely used in petroleum industry. Some of the most important applications in petroleum industry using surfactants in petroleum recovery and processing industry, oilwell drilling, reservoir injection, oilwell production and seagoing transportation of petroleum emulsions [30]. Before speaking about the most important applications in oil industry, a brief account on the chemical composition of crude oil should be given.

1.5.1. Petroleum crude oil composition

Petroleum means rock oil. It is natural organic material occurs in the gaseous or liquid state. The liquid part obtained after the removal of the dissolved gas is commonly referred to as crude petroleum or crude oil [80]. Crude oil is a complex of hydrocarbons with small amounts of sulfur, oxygen and nitrogen, as well as various metallic constituents, particularly vanadium, nickel, iron and copper [81]. Generally crude oil can be separated into some major fractions such as saturates, aromatics, resins, and asphaltenes.

Changes in the balancing of the crude oil phases during production and processing may lead to the formation of solid phases through the precipitation of heavy fractions, such as asphaltenes, resins and paraffin. Due to their aggregative nature , also, acids and heavy paraffins may coprecipitate with asphaltenes [82]. Sediments can also be formed from organic materials but the usual interference is that these materials are usually formed from inorganic materials. The inorganic materials can be salt, sand, rust, and other contaminants that are insoluble in the crude oil and which settle to the bottom of the storage vessel. Tank bottom sludge was characterized as having higher concentrations of lead, zinc and mercury, but lower concentrations of nickel, copper and chromium.

The instability or incompatibility of crude oil is manifested in the formation of sludge, sedment, and general darkening in color of the liquid. Sludge takes one of the following forms: (1) material dissolved in the liquid; (2) precipitated material; and (3) material emulsified in the liquid, sludge which is not soluble in the crude oil may either settle at the bottom of the storage tanks or remaining the crude oil as an emulsion. The existent dry sludge is operationally defined as the material separated from the bulk of a crude oil or crude oil product by filteration and which is insoluble in heptane.

Asphaltene and resins are important deposit forming crude oils heavy fractions [83]. Both asphaltenes and waxes increased the interfacial tension while aromatics and resins decreased it [84]. A better knowledge of the composition of the deposit plus a rheological study is necessary in order to evaluate the best effective and economical treatment. Problems related to the crystallization and deposition of heavy organic fractions during production, transportation and storage of crude oils can lead to considerable financial losses for the petroleum industry. The heavy organic fractions may include paraffins or waxes, resins, asphaltenes, and organometallic compounds that exist in crude oils in various quantities, states, and forms. The severity of the deposition problems varies widely and depends on factors such as the crude oil composition, these composition are:

a) Asphaltenes

Asphaltenes precipitation is one of the most undesirable situations in crude oil production and processing, causing severe technical problems include the formation of organic deposits in oil reservoirs; wells, transport pipelines and equipment and significant economic losses. In order to effectively control asphaltene deposition, the nature and behavior of asphaltene in crude oils is complex and potential solutions for these problems include physical removal of deposits, solvent washes and treatment with dispersant agents [85]. Asphaltenes are defined according to their solubility in non-polar solvents and not by their chemical structure. The common definition is that they are the crude oil components insoluble in n-heptane and soluble in toluene or benzene [86]. However, other definitions are based on the use of a different precipitating solvent, such as n-hexane and *n*-pentane [87]. An important aspect in the preparation of asphaltenes is the proportion of crude oil/n-heptane used to precipitate them. Some investigators recently, a crude oil/n-pentane ratio of 3:2 has also been used to precipitate asphaltenes [88].

Asphaltenes also play a role in the rheological properties of asphalt [81]. The chemical structure and physicochemical properties of asphaltenes are not well understood. However, ^{1}H-NMR and infrared-spectroscopic data showed that asphaltene molecules contain condensed polynuclear aromatic rings with alkyl side chains and heteroatoms such as N, O, S and Ni [89]. The polarity of asphaltenes has often been related to heteroatom or metal content [90], studied qualitatively the polarity of asphaltenes, resins, and oils by dielectric constant measurements in benzene solution at different frequencies. Asphaltenes were found to be highly polar while resins had intermediate polarity between asphaltenes and oils.

Recent evidence suggests that asphaltenes are macromolecules that self-associate in a manner analogous to polymerization [91]. X-ray diffraction experiments suggest that the structure of asphaltene can be represented by flat sheets of condensed aromatic systems interconnected by sulfide, ether, aliphatic chains or naphthenic-ring linkages. However, some recent x-ray and neutron-scattering results suggest that asphaltene particles appear to be spherical with average radii in the range 3-6 nm [92]. Rheological and ultramicroscopic studies also suggest an essentially spherical shape of asphaltene particles in crude oil [93]. On the other hand, the molecular weight (MW) of asphaltenes has been a matter of controversy for more than two decades. Differences in reported average MWs reach a factor of two or more. Groenzin and Mullins's [87] results, obtained from different methods, range from 700 to 4000 amu. Other authors [94] using gel permeation chromatography (GPC), have found that the average MW of asphaltene is as high as 9500 amu. Complex composition and chemical association make it difficult to measure accurately the molecular weights of asphaltenes, association between asphaltene and resin molecules provides further complications. As a result, a wide range of asphaltene molecular weights has been reported (500-50,000 Da). However, data in highly polar solvents (such as pyridine) indicate that, although the molecular weights of asphaltenes are variable, they usually fall into the range 2,000 ± 500 Da [95]. Even lower asphaltene molecular weights are observed in hot, polar solvents suggesting that aggregates formed through stacking interactions between asphaltene monomers. The most mechanisms of asphaltene aggregation involve π/π overlap between aromatic sheets, hydrogen bonding between functional groups and other charge transfer interactions [96].

The aliphatic chain and the polynuclear aromatic core of an asphaltene molecule represent the solvent-hating and the solvent-loving portions, respectively, which may cause asphaltene micelle formation. They also indicate that an increase in the concentration of asphaltene in toluene will result in the formation of larger micelles, then of coacervates and finally, to the separation of asphaltene as a separate phase. The possibility of coacervation of asphaltene micelles were originally reported [97]. Experimental data for the micelle coacervation point of asphaltene through viscosity measurements were produced [98]. Scattering techniques have shown that the best model to describe the morphology of asphaltene micelle in solution is a disk. However, several experimental investigations have indicated that asphaltene micelles could be of spherical-like, cylindrical-like or disk-like form [99] and show in **Figure (1.2)**.

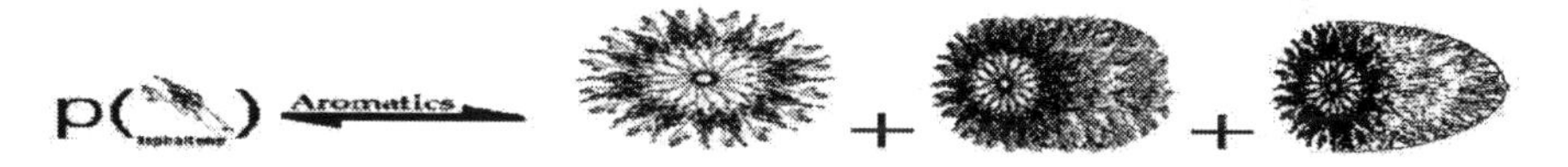

Figure (1.2): Forms of asphaltene micelles

All these investigations are indicative of the fact that asphaltene particles may self-associate, but not flocculate, and form micelles in the presence aromatic hydrocarbons (or other polar solvents). An example of asphaltene micelle coacervation is depicted in **Figure (1.3)**.

There is much evidence to suggest that asphaltenes in crude oil are stabilized by resins with properties similar to those of asphaltenes, but with lower molecular weight and lower polarity [93]. It has been shown that without a resin fraction, asphaltenes cannot dissolve in crude oil. In addition, resin-asphaltene interactions appear to be preferred over asphaltene-asphaltene interactions [100]. The layers of resin on large asphaltene particles will then repel each other if they are in a ''good'' solvent, and this overcomes the van der Waals attraction so that the asphaltene particles will not aggregate. Therefore, we expected that asphaltenes in crude oil exist as single molecules or small aggregates of asphaltene molecules peptized by resin molecules.

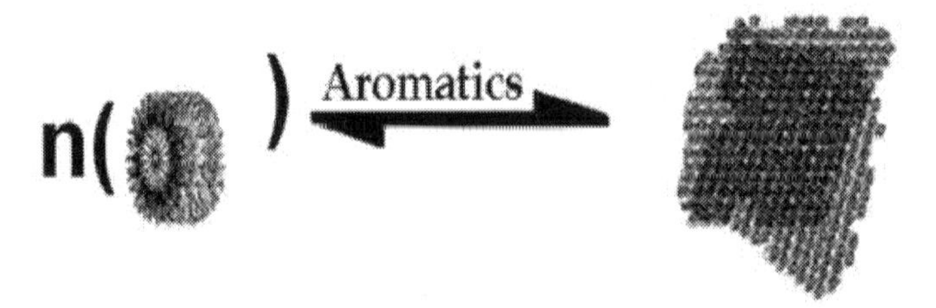

Figure (1.3): Coacervation (self-association) of asphaltene micelles due to increase of micelle concentration in an aromatic (polar) medium

Asphaltenes and resins are polar and may associate to form micelles, in a petroleum fluid, asphaltenes and resins exist in the form of monomers, as well as in micelles [93]. In a micelle, the micellar core is formed from the self-association of asphaltene molecules and resins adsorb onto the core surface to form the shell that also contains the oils [101] as shown in **Figure (1.4)**.

In a crude oil/water system, the asphaltene molecules behave like a surfactant and adsorb onto the surface of water droplets to form water-in-oil emulsions. However, one cannot divide an asphaltene molecule into a polar head and a nonpolar tail as surfactants such as amphiphiles do. In a crude oil, asphaltene molecules do not act as a surfactant, but resin molecules do.

Asphaltenes can be peptized in normal-alkane solvents by amphiphile molecules such as p-(n-dodecyl) benzene sulfonic acid (DBSA), alkylamines and nonyl phenol. As mentioned in [102], it was found that the ability of an amphiphile to stabilize asphaltenes depends on the polarity of its head group and on the length of the alkyl tail which make a steric alkyl layer around asphaltene molecules; they suggested that the acid-base interactions between asphaltenes and natural resins may account for the stabilization of asphaltenes in crude oil [103].

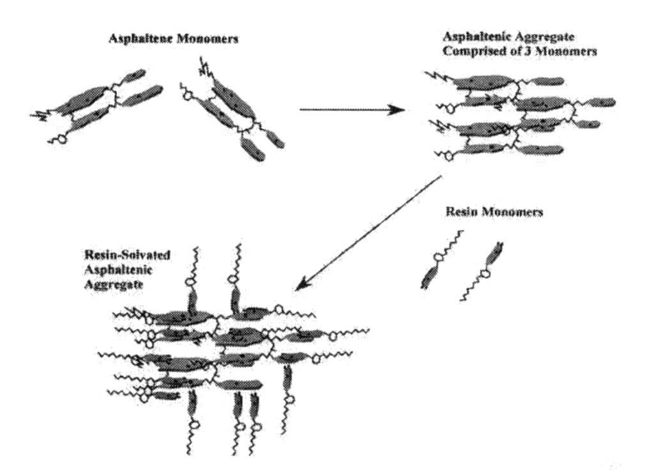

Figure (1.4): Schematic illustration the model of asphaltene monomers, asphaltene aggregate in the absence of resins and asphaltene aggregate in presence of resins.

b) Resins

Resins are defined as the fraction of the desasphalted oil that is strongly adsorbed to surface-active materials such as Fuller's earth, alumina, or silica, and can only be desorbed by a solvent such as pyridine or a mixture of toluene and methanol. Asphaltenes and resins are aromatic heterocompounds with aliphatic substitutions and they form the most polar fraction of crude oil. Resins have a strong tendency to associate with asphaltenes. Such association determines, to a large extent, their solubility in crude oil [104]. The molecular structure of resin has received less attention than that of asphaltene. It has been postulated that resin molecules contain long paraffinic chains with naphthenic rings and polar groups (such as hydroxyl groups, acid or ester functions) interspersed throughout. The molecular weights of resins are about 800 daltons, substantially lower than those of asphaltenes. Unlike those for asphaltenes, the measured molecular weights of resins do not usually vary with the nature of the solvent or temperature. Therefore, chemical association between resin molecules is unlikely

[89]. Resin also indicates the presence of hydrogen bonded hydroxyl groups and another band to N-H functions in pyrroles or indoles.

Another main objective is to find how to measure the amount of resins in a petroleum fluid. A petroleum fluid is a continuum of several thousands of molecules, and it is very difficult to define a cut off between the asphaltene, the resin, and the oil fractions. Nevertheless, asphaltene molecules are believed to contain one or two aromatic chromophores (containing on average seven rings) with short aliphatic side-chains [105]. Resin molecules have a similar structure but smaller chromophores and relatively longer aliphatic side-chains which increase their solubility in aliphatic solvents [101].

c) Paraffins

n-Paraffins or waxes are composed of long chain alkanes with the general formula $C_n H_{2n+2}$, they are characterized by a relatively high melting point when compared to liquid hydrocarbons [106]. There are other classes of paraffins: isoparaffins which are saturated branched chain hydrocarbons and cycloparaffins which have saturated rings with alkyl side chains. The iso- and cycloparaffins have lower melting points and greater solubility in the oil. Therefore, they usually remain dissolved until it reaches lower temperatures. When they do deposite, their configuration does not permit the stronger crystal structure built by *n*-paraffins.

The problems manifested themselves through the accumulation of solid deposits in the storage tanks at a terminal and originated from the mixtures of various crude oils transported to the terminal by pipeline. A small content of 2-3% of crystallized paraffins is able to form a solid interlocking network of fine sheets. And the presence of high molecular weight hydrocarbons (HMWHCs) and more specifically those above C35, in a crude oil may cause major economic problems if deposition of these HMWHCs is allowed to occur during change conditions in the storage tanks at the shipping terminal.

In addition to the HMWHCs, the solid deposits may also contain resins, gums, asphaltenes, oil, water, and inorganic matter [107].

The pour point of a crude oil is related to the temperature at which the oil gels due to the crystallisation of waxes as platelets or needles, forming a three-dimensional network leading to solidification of the crude. Wax from $C_{18}H_{30}$ to $C_{40}H_{82}$) mixed with other organic and inorganic materials, deposits consist of straight and branched chain hydrocarbons (usually ranging) [108].

1.5.2. Surfactant as demulsifier for petrolium crude oil

Emulsions are not always desirable. In the petroleum industry, they can result in high costs and extensive damage to the environment. Emulsions often possess very complex behavior, demulsifiers are a class of surfactants used to destabilize emulsions and this destabilization is achieved by reducing the interfacial tension at the emulsion interface often by neutralizing the effect of other or counteracts the effect of surfactants naturally present in the wellhead or process emulsion. To optimize demulsifier selection for a particular emulsion, it should be sufficient, in principle, to obtain a complete chemical and physical characterization of both the emulsion to be separated and the demulsifier to be used. Many attempts have been made to arrive at some correlation between demulsifier performance and properties like HLB, partition coefficient, interfacial viscosity, and interfacialtension [109]. The factors that can affect on demulsifier performanc are Oil type, oil viscosity, and the size distribution of the dispersed water phase, hydrophile–lipophile balance (HLB), and interfacial tension [110]. Finely divided solids in contact with oil and water can also form solid-stabilized emulsions. These fine particles adsorb at the droplet surface and, by lowering the demulsification rate constant, act as a barrier to prevent droplet coalescence [111]. Oil is produced in combination with water as an emulsion. Some fraction of the water separates easily, while a portion is emulsified and requires chemical or mechanical processing.

1.5.2.1 Petroleum crude oil emulsions

A crude oil emulsion is a dispersion of water droplets in oil. Produced oil field emulsions can be classified into three broad groups [112].

Water-in-oil (W/O) emulsions. In which water droplets are in a continuous oil phase. Generally W/O emulsions are more common in petroleum industry because most produced oil field emulsions are of this kind.

Oil-in-water (O/W) emulsions. This kind of emulsions consists of oil droplets in a continuous water phase.

Multiple or complex emulsions (W/O/W). The resulted or formed emulsions in this case are more complex and consist of tiny droplets suspended in bigger droplets that are suspended in a continuous phase. For example, a water-in-oil-in-water (W/O/W) emulsion consists of water droplets that in turn suspended in a continuous water phase.

Oil field emulsions are characterizes by a number of properties including appearance, basic sediment and water, droplet size, bulk and interfacial viscosities, and conductivities. A brief description of some of these properties is given below [113].

Droplet- size and droplet-size distribution. Produced oil field emulsions generally have droplet diameters exceeding 0.1μm and may be larger than 50 μm. The droplet-size distribution in an emulsion depends on a number of factors, including nature of emulsifying agent, presence of solids and bulk properties of oil and water. Droplet-size distribution in an emulsion determines to a certain extent the stability of the emulsion and should be taken into consideration in the selection of the optimum-treatment protocols.

Viscosity of emulsions. The viscosity of an emulsion can be substantially higher than the viscosity of either the oil or the water. The viscosity of emulsions depends on a number of factors, Such as:-

- Viscosities of oil and water.
- Volume fraction of water dispersed.
- Droplet-size distribution.

- Temperature.
- Shear rate and amount of solids present.

Emulsion viscosity is measured by standard viscometers and rheometers, such as capillary-tube and rotational viscometers.

1.5.2.2 Crude oil emulsion stabilized by natural surfactant

About 80% of exploited crude oils exist in an emulsion state, all over the world. The more common emulsions in the petroleum industry are of the water-in-oil type. Naturally occurring surfactants in crude oils have been identified as largely responsible for the stability of these emulsions. The surfactants that exist naturally, such as asphaltenes and resins, apparently promote the stability of the emulsions by forming highly viscous or rigid films at the oil-water interface [4]. The films have two characteristics, first, interfacial tension is some what high; in other words, the possibility of interfacial activity of the naturally occurring surfactants in forming films is not high, second, the strength of the films is high; that is the principal source of the stability of the emulsions is given rise to by the formation of condensed and viscous interfacial films at the oil–water interface. The materials that form the films are mainly asphaltenes, resins, waxes, petroleum acid soaps, and clay particles. McLean et al. reported that the emulsion stability is governed primarily by the solubility of the asphaltenes in the crude oil [114]. The characteristics of the crude oil, which should play a role in determining the solubility of the asphaltenes, include the resin-to-asphaltene ratio, the aromaticity ratio of the crude medium and resins with respect to the asphaltenes, and the concentration of polar functional groups in the resins and asphaltenes. Zaki et al. studied the effect of asphaltenes, resins, and paraffin waxes on the stability of water-in-waxy oil emulsions [115]. A 20 wt% paraffin wax dissolved in decalin was used to model the waxy crude oil. They indicated that resins alone were not capable of stabilizing the emulsions. Therefore, it is of interest to isolate and characterize these specific crude oil components. For economic and operational reasons, it is necessary to separate the water completely from the crude oils before transporting or refining them. Minimizing the water levels in the oils can reduce pipeline corrosion and maximize pipeline usage [116]. During the process of oil extraction, produced water (oily wastewater) is generated after dehydration of the produced liquid, which is a mixture of oil and water and pumped directly from oil wells in oilfields [117, 118]. Thus water in oil emulsions form spontaneously during oil exploitation. These emulsions are strongly stabilized by native crude oil surfactants which prevent droplet coalescence by forming a rigid film. Asphaltenes, resins, waxes and small solid particles are generally considered to be responsible for the emulsion stability, but the mechanism is not fully understood [119]. Many studies were performed on the film forming and emulsifying behavior of crude oil- water systems. These and later studies often pointed to the asphaltenic constituents in crude oil as being responsible for film formation and stabilization [120]. Asphaltene chemistry plays a vital role in dictating emulsion stability. The most polar species typically required significantly higher resin concentrations to disrupt asphaltene interactions and completely destabilize emulsions. Aggregation and film formation are likely driven by polar heteroatom interactions, such as hydrogen bonding, which allow asphaltenes to absorb, consolidate, and form cohesive films at the oil- water interface [121]. Aspartames and resins both have a largely hydrophobic hydrocarbon structure containing some hydrophilic functional groups and consequently are surface-active [4, 122]. The form in which asphaltenes and resins adsorb on the emulsion interface depends on the form at which they exist in solution. However, the behavior of asphaltenes and resins in crude oil is still poorly understood. Asphaltenes are known to self-associate in crude oil and form colloids or micelles [123]. Resins are known to strongly inhibit asphaltene association [124]. The resin molecules may adsorb on the asphaltene aggregates [122, 123] or may simply act as a good solvent. Therefore, asphaltenes and resins may adsorb on the emulsion interface as independent molecules or as different types of aggregates. Eley *et al* [125] proposed that asphaltenes stabilize water-in-oil emulsions if they are near or above the point of incipient flocculation; that is, they may be solid particles. Others have suggested that asphaltene colloids are responsible for stable emulsions [126]. The asphaltenes may collect on the interface in the form of fine solid particles or asphaltene–resin colloids. The examination of the interfacial films formed by means of the Langmuir–Blodgett technique and the thin liquid film pressure balance technique indicated that asphaltenes adsorb on the interface in molecular rather than colloidal form [127]. It is of interest to determine the interfacial composition at higher asphaltene concentrations that are closer to realistic crude oil compositions. It is also necessary to consider the effect of resins as they may influence asphaltene aggregation, adsorption on the interface, and emulsion stability [5].

Stabilization provided by polymeric surfactants has been found to have numerous advantages. For example, sterically stabilized systems are much less sensitive to fluctuations and increases in electrolyte concentrations, they are equally effective at high and low solids contents and most of all and they stabilize aqueous and nonaqueous dispersions equally well. In the last couple of decades a number of research scientists have synthesized a variety of interesting polymeric amphipathic materials for the use as stabilizer in aqueous and nonaqueous dispersions. Besides the use of polymeric surfactants as dispersion stabilizers, other important and novel applications of these materials have been developed. They have been utilized as dispersion thickeners for rheology control as hydrogels in microemulsions and in numerous other capacities [23].

1.5.2.3. Mechanisms of demulsifier action

Demulsification is the separation of an emulsion into its component phases. It is a two-step process. The first step is flocculation (or aggregation, agglomeration, coagulation). The second is coalescence. Either of these steps can be the rate-determing step in emulsion-breaking process.

Flocculation or Aggregation. The first step in demulsification process is flocculation of water droplets. During flocculation, the droplets clump together forming aggregates, or "flocs". The droplets are close to each other, even touching in certain points but may not lose their identity (e.g., they may not coalesce). Coalescence at this stage takes place only if the interfacial film surrounding the water droplets is very weak. The rate of flocculation depends on a number of factors including water cut, temperature, viscosity of the oil, and the density difference between the oil and water.

Coalescence. It is the second step in the demulsification process and follows flocculation. During coalescence, water droplets fuse, or coalesce to form a larger droplet. This is an irreversible process that leads to a decrease in the number of water droplets, and eventually to complete demulsification. Coalescence is enhanced by a high rate of flocculation, absence of mechanically strong films, low oil and interfacial viscosities, high IFTs, high water cuts and high temperatures [128].

Demulsification is the breaking of a crude-oil emulsion into oil and water phases. Demulsifiers are a class of surfactants used to destabilize emulsions. This destabilization is achieved by reducing the interfacial tension at the emulsion interface, often by neutralizing the effect of other naturally occurring surfactants which are stabilizing the emulsion. Produced oilfield emulsions posses a degree of kinetic stability arises from the formation of interfacial films encapsulating the water droplets. Therefore, destabilizing or breaking emulsions is linked very intimately to the removal of these interfacial films. Demulsification by use of chemicals or demulsifiers is a very complex phenomenon. Demulsifiers displace the natural stabilizers (emulsifiers) present in the interfacial film around the water droplets. This displacement is brought about by the adsorption of the demulsifier at the interface. This displacement occurring at the oil / water interface, influence the coalescence of water droplets through enhanced film drainage. The efficiency of the demulsifier is dependent on its adsorption at the oil / water, or droplet surface [129].

As mentioned previously, produced oilfield emulsions are stabilized by films that form around the water droplets at the oil / water interface (interfacial films). These films are believed to result from the adsorption of high-molecular-weight polar molecules that are interfacially active (i.e., exhibit surfactant-like behavior). These films enhance the stability of emulsions by reducing IFT and increasing the interfacial viscosity. The characteristics of interfacial films are a function of the crude-oil type (*e.g.*, asphaltic or paraffinic), composition and pH of water, temperature, the extent to which the adsorbed film is compressed, contact or aging time, and concentration of polar molecules in the crude oil. These films are classified into two categories on the bases of their mobilities [130].

Rigid or solid films. These are like an insoluble skin on water droplets and are characterized by very-high interfacial viscosity. They are formed by polar fractions of the oil and other emulsifiers. They provide a structural barrier to droplet coalescence and increases emulsion stability.

Mobile or liquid films. These films are mobile and characterized by low interfacial viscosities. They are formed when a demulsifier is added to an emulsion and coalescence of water droplets is enhanced.

Polymeric surfactants are useful mainly as demulsifiers, dispersing agents, and emulsion stabilizers. Their wetting, foam formation, and detergency are much inferior to those of low molecular weight surfactants due to low effectiveness in surface tension reduction. It has been found that many oil-in-water emulsions can only be effectively broken with clean resolution of discontinuous and continuous phases, by treatment with large molecules such as long chain polymeric materials, and the present invention utilizes a novel class of water soluble, water-in-oil emulsions cationic polymeric materials derived from acrylamide and quaternary compound of diallylamine [131].

Many studies were carried for synthesis of polymeric surfactants for treating crude oil emulsions [17, 132]. The first anionic surfactants used as demulsifiers are known as soaps and are usually prepared by saponification of natural fatty acid glycerides in alkaline solution. The degree of water solubility is controlled by the length of the alkyl chain ranging from 12 to 18. The cationic agents are long-chain cations, such as amine salts and quaternary ammonium salts. The nonionic agents offer advantages regarding compatibility, stability and efficiency compared to the anionic and cationic agents. The ethylene oxide and propylene oxide block copolymers are a class of molecules that are particulary active at the oil / water interface. The low molecular weight demulsifiers can be transformed into high-molecular-weight products by reactions with diacida, diepoxides, diisocyanates, and aldehydes. These demulsifiers have the ability to adhere to natural substances that stabilize emulsions such as asphaltenes and resins. These properties increase the demulsification efficiency of such materials [30].

CHAPTER II

"RESULTS AND DISCUSSION"

Water-in-oil (W/O) emulsions are very common in the petroleum industry. They form naturally during crude oil production and the water content can be as high as 60% in volume. Such emulsions can also form accidentally during the refining operations, storage and distribution [133]. It is well known that the formation of these emulsions is responsible for several economic and environmental problems. Demulsification is the breaking of a crude-oil emulsion into oil and water phases. Demulsifiers are a class of surfactants used to destabilize emulsions. This destabilization is achieved by reducing the interfacial tension at the emulsion interface, often by neutralizing the effect of other naturally occurring surfactants which are stabilizing the emulsion. Produced oilfield emulsions posses a degree of kinetic stability arises from the formation of interfacial films encapsulating the water droplets. Therefore, destabilizing or breaking emulsions is linked very intimately to the removal of these interfacial films [134].

The choice of the surfactants (emulsifiers, demulsifiers and corrosion inhibitors) is based on two considerations: first it could be synthesized conveniently from relatively cheap raw materials; secondly, it contains both hydrophilic and hydrophobic moieties. From the above reasons, this study aims to prepare poly(oxyethylene)-co-poly(oxypropylene), PEO-PPO-PEO, block copolymers to use them as demulsifiers. It is well known that PEO-PPO-PEO block copolymers were prepared by anionic polymerization and they required pressure reactor to complete polymerization. The present work aims to prepare this copolymer in the normal condition. Evaluation of the prepared surfactants as demulsifiers for crude oil emulsions is the main goal of the present study.

2.1. Preparation of PEG-PPO-PEG Co-polymers

The first step in this work is the polymerization of propylene glycol and propylene oxide using KOH as a catalyst to produce poly (propylene oxide), PPO1, at the reaction temperature 298 K. The reaction time was adjusted to produce PPO having different molecular weights. The number-average molar mass (Mn), weight-average molar mass (Mw) and polydispersity (PD) for prepared PPO1 are reported as 416, 1348 g/mol and 3.24, respectively **Figure (2.1)**. While, Mn, Mw and PD values for PPO2 are recorded as 3149, 5215 g/mol and 1.65, respectively **Figure (2.2)**. The data indicated that the PD values were decreased with increasing the molecular weights of PPO. The retention time of PPO1 and PPO2 was determined at 29.5 and 28.36 minute, respectively.

Considerable interest has been developed in polymeric surfactant system both as models for membrane mimetic chemistry and for the practical uses, e.g., in process of enhanced oil recovery, emulsifiers, dispersants, demulsifiers and corrosion inhibitors. In this respect, poly(oxyethylene)-co-poly(oxypropylene) polymeric surfactants were prepared by two methods. In the first method PPO was esterified with MA to produce poly (propylene oxide) maleate diester. The produced polymers were esterified with PEG having different molecular weights to produce PEG-PPO-PEG block copolymers. In this respect, PPO1 maleate diester was reacted with PEG 400 and PEG600. PPO2 ester was reacted with PEG 600, PEG 1500 and PEG3000 to study the effect of both chemical structure and molecular weight of hydrophilic and hydrophobic parts of the prepared polymers on surface activities of the prepared surfactants. The produced polymers can be designed as PEG400-PPO1-PEG400, PEG600-PPO1-PEG600, PEG600-PPO2-PEG600, PEG1500-PPO2-PEG1500 and PEG3000-PPO2-PEG3000, respectively (**Scheme 2.1**).

The molecular weights of the prepared PEG-PPO-PEG block copolymers were determined by GPC technique. The resultant molecular weights are listed in Table (2.1) in conjunction with the acid number and hydroxyl values of the prepared PEG-PPO-PEG block copolymers. End group in condensation polymers usually involves chemical methods of analysis for functional groups. Carboxyl and hydroxyl groups in polyester are usually titrated with a titrable reagent. The degree of polymerization of the prepared PEG-PPO-PEG block copolymers calculated from molecular weight of PEG-PPO-PEG block copolymers, which was calculated from hydroxyl number and acid value, and are listed in Table (2.1). Careful inspection of data indicates that the molecular weights of PEG-PPO-PEG block copolymers determined from both end group analysis and GPC technique have the same values and they agree with the theoretical values.

CH_3 CH(OH) CH_2(OH) + CH_3 CH(O)CH_2 —(OH^-, 25 °C)→ HO [CH_2 - CHO(CH_3)]$_n$ H

PG PO PPO

HO [CH_2 - CHO(CH_3)]$_n$ H + MA —($-H_2O$, PTSA)→

PPO MA

HOCO-CH=CH -COO [CH_2 - CHO(CH_3)]$_n$ CO-CH=CH-COOH

I

I + HO [CH_2 - CH_2O]$_n$ H —($- H_2O$, PTSA)→

PEG

HO - [CH_2 - CH_2O]$_n$ -CO-CH=CH -CO O [CH_2 - CHO(CH_3)]$_n$ CO-CH=CH-COO [CH_2 - CH_2O]$_n$ H

PEG-PPO-PEG

Scheme (2.1): Preparation of PEG-PPO-PEG block copolymers

The chemical structure of the prepared PEG-PPO-PEG block copolymers was elucidated by IR and ^{1}H-NMR analyses. In this respect, the IR spectrum of PEG600-PPO1-PEG600 was selected as representative sample and represented in **Figure (2.3)**. Characteristic peaks of oxyethyl and oxypropyl in non-ionic surfactant is 1200 cm^{-1} (υ_{as} C-O-C), strong peak. Series of weak bands were observed in 3450 cm^{-1} (υ_{as} OH, wide)

[135]. The absorption bands from 2850 to 3000 cm^{-1} are saturated C- H stretching vibrations. The absorptions at 2968.6 and 2867.9 cm^{-1} are assigned as υ_{as} and υ_s of the CH_3 group. The absorptions at 2925 and 2850 cm^{-1} are assigned as υ_{as} and υ_s of the CH_2 group. Two absorption bands at 1457.3 and 1373.4 cm^{-1} are methyl group; the CH_2 group has absorption at 1347.4 cm^{-1}. The absorption of C= C in MA is 1650 cm^{-1}, a wide peak was observed in all spectra of PEG-PPO-PEG block copolymers. The appearance of strong bands at 1735 cm^{-1} (υ_{as} C=O ester) and 1100 cm^{-1} (υ_{as} C-O ester) indicated the formation of maleate diester in the chemical structure of PEG-PPO-PEG block copolymers.

Table (2.1): Molecular weights data of PEG-PPO-PEG block copolymers

Designation	GPC data			End group analysis data		
	Mn (g/mol)	Mw (g/mol)	PD	A (acid value) mg KOH/g	B Hydroxyl number mg KOH/g	Mn (g/mol)
PEG400-PPO1-PEG400	1440	3600	2.5	1.3	80.0	1380
PEG600-PPO1-PEG600	1980	6100	3.0	1.1	60.6	1790
PEG600-PPO2-PEG600	4700	7100	1.5	1.3	23.6	4500
PEG1500-PPO2-PEG1500	6400	10240	1.6	1.5	15.1	6300
PEG3000-PPO2-PEG3000	9230	15980	1.7	1.3	10.5	9530

The ^{1}H-NMR spectroscopic analysis was used previously for determining the propylene oxide/ethylene oxide ratio for the PO-EO block copolymers [133, 134, 136]. The chemical structure of PEG-PPO-PEG block copolymers was elucidated by ^{1}H-NMR analysis. In this respect, ^{1}H-NMR spectrum of PEG600-PPO2-PEG600 was represented in **Figure (2.4)**. According to ^{1}H-NMRspectra the chemical shifts from 3.5 to 4.2 ppm are the chemical shifts of methylene group and methine group connected with oxygen and ester groups of maleate ester. The peak at 0.8 ppm is attributed to methyl group of oxypropylene group. The peak at 5.8 ppm can be attributed to double bond of maleate ester group of PEG-PPO-PEG block copolymers. The IR and ^{1}H-NMR spectra of some of the prepared PEG-PPO-PPG block copolymers are shown in **Figures (2.5.a, b)** & **(2.6.a-d)**, respectively.

2.2. Preparation of PPO-g-PEG GRAFT Copolymers

The second method based on the acetylation reaction of PPO2 by acetic anhydride to produce PPO diacetate (PPO-Ac). The produced polymers were esterified with different weight ratio of MA. The produced polymers can were esterfied with PEGME400 and PEGME600 to study the effect of both chemical structure and molecular weight of hydrophilic and hydrophobic parts of the prepared polymers on surface activities of the prepared surfactants. The produced polymers can be designated as PPO-MA(5)-PEG400, PPO-MA(5)-PEG600, PPO-MA(10)-PEG400, PPO-MA(10)-PEG600, PPO-MA(20)-PEG400 and PPO-MA(20)-PEG600, respectively (**Scheme 2.2**). Poly (propylene oxide) is a widely applied type of aliphatic polyether was used for the preparation of surfactants. For the development of advanced water soluble surfactants, a well-defined polar modification and functionalization of PPO could be important. To realize this goal, Frentzel et al [137] and O'Connor et al. [138] have carried out the grafting of PPO with maleic acid and fumaric acid. Taking into account the higher reactivity of maleic anhydride at radical graft reactions compared to maleic acid and fumaric acid, a more efficient grafting can be expected by using MA as graft monomer. Therefore, MA has been used frequently as monomer for the functionalization of polyolefines via radical graft reactions. However, different structures of grafted products have been discussed. Gaylord [139, 140] proposed the formation of poly(maleic anhydride) side chains by the grafting of MA onto polyethylene whereas Russell and Kelusky [141] described the formation of monosubstituted succinic anhydride units by graft reactions of MA onto eicosane. For the grafting of MA onto polypropylene some authors suggest the formation of single succinic anhydride units as well [142, 143]. In contrast [144] the formation of disubstituted succinic anhydride units were considered. De Roover [144] proposed the formation of poly(maleic anhydride) units attached to the chain-ends of polypropylene. Heinen and co-workers [145] found by means of ^{13}CNMR spectroscopy that grafting onto the methylenc units of polyethylene occurs as both single and oligo units, whereas grafting onto the methine units of polypropylene or ethylene/propylene copolymers occurs only as monosubstituted units. A defined structure modification of the aliphatic polyether poly(tetrahydrofuran) via grafting with MAn has been described [146]. In contrast to grafting reactions onto polyolefines, unusual high grafting values up to 20 wt.-% were obtained by the grafting of MA onto poly(tetrahydrofuran). A characterization of the structure of the grafted products evidenced the formation of monosubstituted succinic anhydrides as graft units attached onto the poly(tetrahydrofuran) carbons in a-position. The present work discusses the functionalizalion and polar modification of PPO by the radical-initiated grafting with MA.

To prevent side-reactions of MA with hydroxyl end groups, PPO was completely acetylated to obtain the diacetate form of PPO (PPO-Ac) prior to the graft reaction as described in **Scheme (2.2)**. **Figure (2.7.a)** shows the carbonyl region of the FTIR spectrum of PPO-Ac. The vibration of acetate end-groups appears at 1738 cm^{-1}. The graft reactions with MA initial concentrations of 5 and 10 wt.-% related to PPO-Ac occurred in homogeneous phase. In contrast, by an MA initial concentration of 20 wt.-% a phase separation with the precipitation of a solid phase took place during the reaction. FTIR spectroscopy is a suiTable method to investigate grafted products of MA and polyolefins [144, 147]. A characteristic shift of the carbonyl valence vibrations of cyclic anhydrides to higher wave numbers was observed due to the graft reactions. **Figure (2.7.b)** depicts the carbonyl region of a reaction product of PPO-Ac and 10 wt. % MA related to PPO-Ac after vacuum distillation. Since unreacted MA was removed completely during this distillation procedure [146] the detected anhydride carbonyl valence vibrations are due to grafted anhydrides. Compared to MA [148], the antisymmetric carbonyl valence vibration of the reaction products is shifted from 1781 cm^{-1} to 1783.5 cm^{-1} and the symmetric vibration is shifted from 1856 cm^{-1} to 1862 cm^{-1}. This finding indicates the formation of a saturated cyclic anhydride structure [147]. The small shift of the carbonyl vibration of acetoxy endgroups to a lower wave number (1734.5 cm^{-1}) is probably caused by the presence of some hydrolyzed anhydride groups of graft units. Vibrations of free carboxyl groups appear below 1730 cm^{-1}.

For a further structure characterization, highly grafted parts of the reaction products were isolated by extracting five times in hexane at room temperature. In this way, three different samples of highly grafted PPO-Ac parts were isolated from the reaction products synthesized using MA initial concentrations of 5, 10 and 20 wt.-% related to PPO-Ac. The solid phase precipitated during the reaction with an initial concentration of 20 wt.-% was taken as fourth sample Table (2.2). The structure of the appropriate samples (I-IV) was investigated. The reaction products as well as the samples of the highly grafted parts (I-IV) were titrated to determine the amount of grafted anhydrides. The conversion of all reaction products was approximately 75% of the initial MA concentration. For example, the product synthesized using an MA initial concentration of 10 wt.-% related to PPO-Ac showed a conversion of 78%. The values of the percentage of grafting were approximately similar for all isolated samples, as depicted in Table (2.2). This indicates that a certain concentration of graft units onto PPO-Ac chains is necessary (approximately 20 wt.-%) to achieve an insolubility of the appropriate chains in hexane. Considering free radical graft reactions of polyolefins, it is known that the reaction of macroradicals causes a crosslinking at polyethylene but a chain scission at polypropylene. However, compared to the GPC curve of PPOAc, the peak of one of the grafted products shows no significant difference of the elution time (**Figure 2.8.a. & b**). Even the chromatogram of the highly grafted parts of the product shows only a slight shift to higher elution times (**Figure 2.8.c**). This can be explained by a higher molecular mass due to the relative high concentration of graft units. However, the depicted chromatograms indicate that neither a chain scission nor a crosslinking takes place during the grafting reaction of MAn onto PPO. This finding is similar to the grafting of MAn onto poly(tetrahydrofuran) [146].

In contrast to graft products of poly(tetrahydrofuran) and MA where only ingle graft units were determined [146], graft products of PPO and MA were found to contain both single and oligo units. This finding indicates, as a matter of principle, the possibility of a homopolymerization of MA under the used reaction condi tions (reaction temperature 353 K, initiator DBP). This possibility is in accordance with results of studies concerning the homopolymerization of MA [149, 150]. However, the discussed dependence of the ether carbon reactivity on stereoelectronic effects of its structure influences the homopolymerization of MA Nakayama et al. [151] obtained oligo(maleic anhydride) by radical reaction in 1,4-dioxane. However, in tetrahydrofuran only the formation of an addition product of MA and tetrahydrofuran occurs [152]. The reason is a smaller activation energy for the hydrogen abstraction from tetrahydrofuran compared to 1,4-dioxane [153, 154].

Despite the fact that in this study the kinetics of the graft reaction were not investigated, the results indicate that the hydrogen abstraction from carbons of poly(tetrahydrofuran) proceeds faster than the hydrogen abstraction from carbons of PPO. This interpretation is supported by the different efficiencies of the two graft reactions. A nearly complete conversion of MA [146] was obtained by grafting of poly(tetrahydrofuran) while under comparable conditions the conversion by grafting onto PPO-Ac was significantly lower (approximately 75%). The grafting occurs onto both methylene and methine carbons at the PPO chain. At graft reactions carried out with low MAn initial concentrations up to 10 wt.-%, monosubstituted succinic anhydrides are formed as graft units. However, at higher MA initial concentrations (20 wt.-%) beside single units a small amount of oligo(maleic anhydride) units are found. The degree of conversion was approximately 75% of the initial MA. However, by extraction with hexane highly grafted PPO chains could be isolated easily from the reaction products. The percentage of grafting of the isolated products was up to 24.3 wt.-

%. The synthesized highly functionalized PPO's are of potential interest as components for crosslinked elastomers. Furthermore, the formed anhydride graft units may be used to carry out further reactions such as esterification with PEG to form new surfactant molecules with new interesting properties.

Table (2.2): Reaction conditions of PPO grafts with MA

Sample	MA concentration Related to PPO-Ac	treatment	Percentage of grafting (Wt %)
PPO-MA5	5	isolated highly grafted parts	22.5
PPO-MA10	10	isolated highly grafted parts	20.4
PPO-MA20	20	isolated highly grafted parts	20.5
PPO-MA20P	20	Precipitated during the reaction	24.5

There are different classes of surfactants derived from maleic anhydride namely the ethoxyalkyl maleates were derived from maleic anhydride. A non-ionic amphiphilic maleic diester with a poly(ethylene oxide) hydrophilic chain was used as a copolymerizable surfactant [155-157]. Surfactants derived from maleic acid have a very important additional advantage: they are not able to homopolymerize. However, as they are polymerizable surfactants, they can form copolymers with the main monomers used in emulsion polymerization. In addition, maleates seem to be rather reactive, with a high level of conversion during the copolymerization process. This makes reactive maleates promising as surfactants for the improvement of surface characteristics of polymer latexes. The first part of the present work aimed to produce nonionic polymeric surfactants from reaction of modified PPO with MA followed by reaction with PEG having different molecular weights to produce PPO block PEG copolymers.

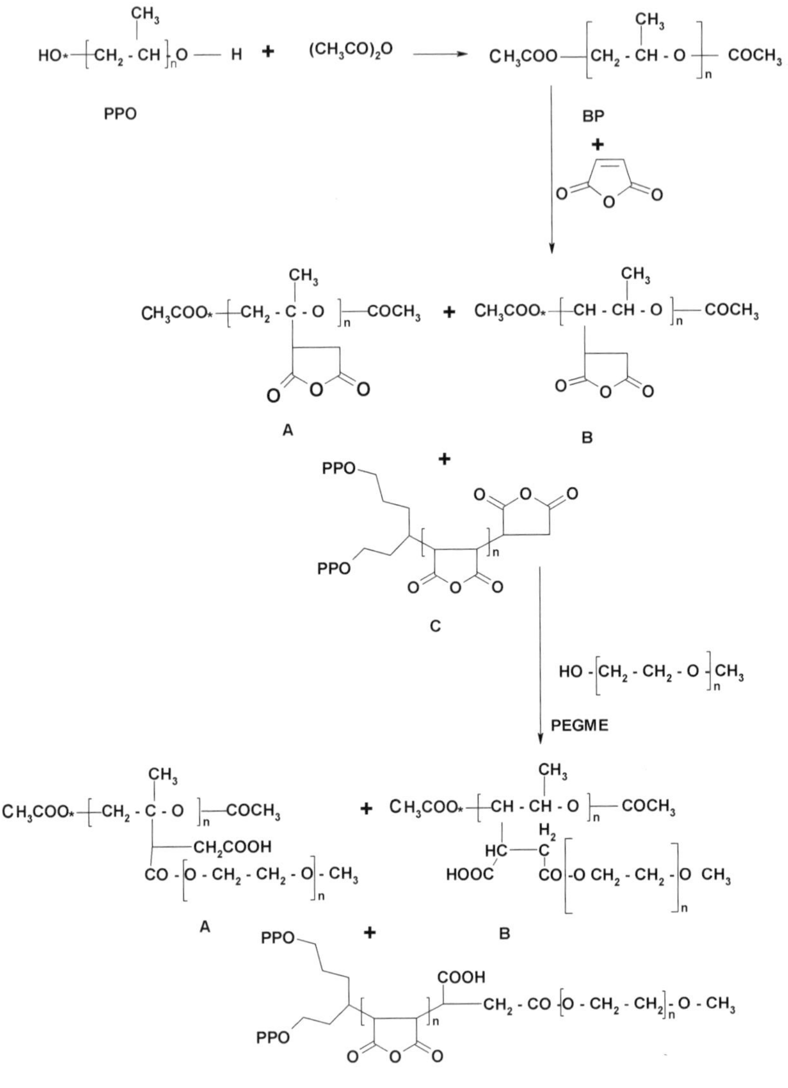

Scheme (2.2): Preparation of PPO-PEG graft copolymers

In the present section PPO-MA grafts were reacted with PEGME to form PPO-g-PPG graft copolymers through formation of PPO-MA grafts followed by esterification with PEGME.

The present work describes the esterification of PPO-MA with PEGME to produce nonionic surfactants having different hydrophile-lipophile balance (HLB) and to study the effect of surfactant structure on its properties. In this respect, the PPO-MA was subjected to react with PEGME having different molecular weights 400 and 600 at 453 K. The produced surfactants are soluble in water, toluene, xylene and $CHCl_3$. Besides that, products with high ester yield can be obtained either by using catalysts or by adding one of the reacting components in large excess or even by removal of water. In view of this, the reaction took place at relatively high temperature in order to remove any water from the reaction medium to shift the reaction to complete esterification. Addition of PTSA significantly increased the reaction rate. The Scheme of reaction with PEGME as model component is given in **Scheme (2.2)**. However basic knowledge of the relation between the structure of the surfactants and their performances is still lacking.

The chemical structures of PPO-MA-PEG were confirmed by IR spectroscopy. The IR spectra of PPO-MA5-PEG600, PPO-MA10-PEG400, PPO-MA10-PEG600 and PPO-MA20-PEG400 were represented in **Figure (2.9.a-d)**. It was observed that the spectra of all the esterified derivatives are nearly identical. Careful inspection to the spectra in **Figure 2.9 (b & c)**, the following bands can be observed; the stretching band at 3450 cm^{-1} for OH of remained unreacted COOH group and the characteristic band for γ C= O in COOH group at 1700 cm^{-1}. The disappearance of two bands of γ C=O of cyclic anhydride group at 1780 and 1810 cm^{-1} and appearance of two characteristic bands of C=O stretching band at 1730 cm^{-1} and C-O of ether at 1110 cm^{-1} indicated the esterification of PPO-MA with PEGME.

Furthermore, the ^{1}H-NMR analysis was used to confirm the structure of the prepared compounds. The ^{1}H-NMR spectra of PPO-MA5-PEG400, PPO-MA10-PEG400, PPO-MA20-PEG400 and PPO-MA20-PEG600 were represented in **Figures (2.10.a-d)**. Careful interpretation to these spectra indicated the appearance of peak at the chemical shift at δ = 3.6 ppm for protons of oxyethylene units, the chemical shift at 4.3 ppm which represented COO-(CH_2CH_2O). These peaks were observed in spectra of all surfactants. There are some other characteristic peaks such as the chemical shift of methyl groups of PPO and PEGME which appeared at δ = 0.92 and 3.2 ppm, respectively. Also, the -OH signal that usually appears at 10.3 ppm in spectrum of MA and assigned for -COOH group appeared in all spectra of PPO-MA-PEGME surfactants. This observation certainly indicates that only one carboxylic group of MA was esterified with PEGME. Moreover, the integration area of terminal COOH signal appeared at 10.3 ppm was compared to the integration area of either –OCH_3 signal, 3.2 ppm, or with the chemical shift at 4.3 ppm which represented COO-(CH_2)- to determine the degree of esterification of PPO-MA with PEGME and to calculate the theoretical molecular weight of the prepared surfactants. Grafting percentages, theoretical molecular weights and esterification precentages of the prepared surfactans were determined and listed in Table (2.3). The data indicated that the percentage of esterification of PEGME was increased with increasing molecular weights of PEGME from 400 to 600 g/mol and with increasing of MA content from 5 to 20 Wt%. This can be attributed to the higher solubility of PPO-MA20 in toluene than PPO-MA5 that enhances the probability for reaction of PPO-MA with PEGME [158].

Table (2.3): Grafting percentages and theoretical molecular weights of PPO-MA-PEGME grafts

Designation	Grafting Precentage[a]		Theoretical Average molecular weights (Mw)[b] g/mol		
	MA (%)	PEGME (%)	PEGME	PPO	PPO-MA-PEGME
PPO-MA5-PEG400	70	20	6151	3149	9300
PPO-MA5-PEG600	70	25	10451	3149	13600
PPO-MA10-PEG400	75	33	9851	3149	13000
PPO-MA10-PEG600	75	38	17051	3149	20200
PPO-MA20-PEG400	78	45	14051	3149	17200
PPO-MA20-PEG600	78	50	23401	3149	26550

[a] Determined from ^{1}H-NMR analyses.

[b] Theoretical M.w. = [MA% x PEGME M.w. x PEGME % / 100] + [PPO M.w.]

2.3. Solubility Properties of the Prepared Surfactants

It is well known that the modification of polymer backbone yields different hydrophobicity, chain flexibility and solubility due to the difference of inter- and intramolecular interactions. Accordingly, the selection of the proper solvent depends to a large extent on the type and quantity of the branches attached to the backbone. This difference in solubility is due to the difference in hydrophil-lipophil balance (HLB) of the surfactants. The HLB values were calculated by using the general formula for nonionic surfactants, HLB = $[M_H / (M_H + M_L)] \times 20$, where M_H is the formula weight of the hydrophilic portion of the surfactant molecule and M_L is the formula weight of the hydrophobic portion. HLB values of nonionic surfactants based on PEG400, PEG600, PEG1500, PEG3000, PEGME400 and PEGME600 were calculated and listed in Table (2.4) and Table (2.5), respectively. It is obvious that the HLB of the surfactants have low values, this can be attributed to the structure of surfactants is more hydrophobic surfactants. HLB is an important parameter characterizing a surfactant that can indicate its appropriate applications. Classical equations derived by Davies were used to calculate the HLB number of surfactants [159], however, these equations consider only the chemical compositions, and the effect of position isomerism is not taken into account. Because HLB is difficult to determine experimentally, we instead used cloud point to represent the hydrophile–lipophile balance. The cloud point is the temperature below which a single phase of molecular or micellar solution exists; above the cloud point the surfactant loses sufficient water solubility and a cloudy dispersion results. Above this temperature, the surfactant also ceases to perform some or all of its normal functions. So cloud point can be used to limit the choice of nonionic surfactants for application in certain processes. A suggestion was made to regard the cloud point in solution of nonionic surfactant as a pseudo-phase inversion. For polyoxyethylene-type surfactant, the cloud point and the phase inversion temperature (PIT) are directly correlated when surfactant alone is dispersed in water. PIT is defined as the temperature at which the hydrophile–lipophile property of surfactant just balances at the interface [160, 161]. As seen from the presented data, the cloud points were progressively decreased with increasing lengths of the hydrophilic side chains and molecular weight of PEG. A study on the effect of structural changes in the surfactant molecule on its cloud point [162, 163] indicates that, at constant oxyethylene content the cloud point is lowered due to the following reasons:

- Decreased molecular weight of the surfactant.
- Broader distribution of polyoxyethylene chain length in commercial materials.
- Branching of the hydrophobic group.
- More central positions of the polyoxyethylene hydrophilic group in the surfactant molecule, replacement of terminal – OH methoxyl.
- Replacement of the ether linkage between the hydrophilic and hydrophobic group by an ester linkage.

Careful inspection of data indicates that the prepared surfactants have different HLB values varied from (5.1-17.1) which indicate that the prepared surfactants have good solubility in both oil and water phases. It is obvious that the HLB values of surfactants of group 1 have lower values than those surfactants of group 2. It has been mentioned that when the surfactants are dissolved in a solvent, materials that contain a lyophobic grouping together with a lyophilic groups in the same molecule distort the structure of the solvent and therefore increase the free energy of the system. On the other hand when they concentrate at the surface and orient their lyophobic groups away from the solvent, the free energy of solution is minimized. However, there are some other means of minimizing the free energy in the systems. The distortion of the solvent structure can also be decreased (and the free energy of the solution reduced) by the aggregation of the surface-active molecules into clusters (micelles) while their hydrophilic groups directed toward the solvent. The micellization is therefore a mechanism alternative to adsorption at the interfaces for removing the lyophobic groups from contact with the solvent, thereby reducing the free energy of the system [160, 162] The data indicated that the best HLB values for dehydration of the crude oil emulsion are ranged from 11.1 to 12.1. This data indicated that PEG400-PPO1-PEG400 and PEG600-PPO1-PEG600 surfactants have good adsorption at oil water interface.

These results agree with the data reported on adsorption parameters and IFT measurements which indicate that PEG400-PPO1-PEG400 and PEG600-PPO1-PEG600 surfactants have best adsorption performance and high reduction in IFT which destabilize crude oil emulsions. This indicates the partial solubility of most of these surfactants in water. The low HLB values of the prepared surfactants will be more soluble in nonpolar solvents [164].The prepared emulsions can be considered as water in oil emulsions, in other words the oil percent in these emulsions is more than water since the prepared surfactants are more soluble in oil than water, so it will be expected to give good dehydration rates with such emulsions.

Table (2.4): The calculated HLB values of the prepared PEG-PPO-PEG surfactants

Surfactants	Molecular weight	HLB	Cloud point (K)
PEG400-PPO1-PEG400	3600	11.1	340-342
PEG600-PPO1-PEG600	6100	12.1	355-357
PEG600-PPO2-PEG600	7100	5.1	331-333
PEG1500-PPO2-PEG1500	10240	9.37	339-342
PEG3000-PPO2-PEG3000	15980	13.0	353-355

Table (2.5): The calculated HLB values of the prepared PPO-MA-PEG surfactants

Surfactants	Molecular weight	HLB	Cloud point (K)
PPO-MA5-PEG400	9300	13.2	353-356
PPO-MA5-PEG600	13600	14.7	348-350
PPO-MA10-PEG400	13000	15.1	351-353
PPO-MA10-PEG600	20200	16.8	346-348
PPO-MA20-PEG400	17200	16.3	343-345
PPO-MA20-PEG600	26550	17.1	338-340

2.4. Surface Activity of the Prepared PEG-PPO-PEG Surfactants

The surface activity of surfactants can be determined by measuring surface or interfacial tensions versus time for a freshly formed surface [30]. The micellization and adsorption of surfactants are based on the critical micelle concentrations (CMC) which was determined by the surface balance method. The CMC values of the prepared polymeric surfactants were determined at 298, 308, 318 and 328K from the change in the slope of the plotted data of surface tension (γ) versus the natural logarithm of the solute concentration. Adsorption isotherms of group 1 (the five prepared PEG-PPO-PEG block copolymers) are shown in **Figure (2.11 a-e)**, while those of group 2 (the six prepared PEG-PPO-PEG block copolymers) are shown in **Figure (2.12 a-f)**, respectively.

The presented plots and all other plots are used for estimating surface activity and confirming the purity of the studied surfactants. It is of interest to mention that all obtained isotherms showed one phase, which is considered as an indication on the purity of the prepared surfactants. It may be possible in these cases that the low water solubility of these surfactants is due to the long alkyl chain, not the EO chain. This behavior is obvious only over some EO ranges, as would be expected from the increase in the hydrophilic character of the molecule resulting from this change. In previous work [165, 166] this behavior was explained on the basis of coiling of the polyethylene oxide chains. The direct determination of the amount of surfactant adsorbed per unit area of liquid-gas or liquid-liquid interface, although possible, is not generally under taken because of the difficulty of isolating the interfacial region from the bulk phase for purpose of analysis when the interfacial region is small, and of measuring the interfacial area when it is large.

Instead, the amount of material adsorbed per unit area of interface is calculated indirectly from surface or interfacial tension measurements. As a result, a plot of surface (or interfacial) tension as a function of equilibrium, concentration of surfactant in one of the liquid phases, rather than an adsorption isotherm, is generally used to describe adsorption of this interface can readily be calculated as surface excess concentration Γ_{max}. The surface excess concentration of surfactant at the interface may therefore be calculated from surface or interfacial tension data by using Equation (2.1).

$$\Gamma_{max} = \frac{1}{RT}\left(\frac{-\partial\gamma}{\partial \ln c}\right)_T \qquad [2.1]$$

Where $\left(\frac{-\partial\gamma}{\partial \ln c}\right)_T$ is the slope of the plot of γ versus ln C at a constant temperature (T), and R is the gas constant in J mol^{-1} K^{-1}. The surface excess concentration at surface saturation is a useful measure of the effectiveness of adsorption of surfactant at the liquid-gas or liquid-liquid interface, since it is the maximum value which adsorption can attain.

The Γ_{max} values were used for calculating the minimum area A_{min} at the aqueous-air interface. The area per molecule at the interface provides information on the degree of the packing and the orientation of the adsorbed surfactant molecules, when compared with the dimensions of the molecule as obtained by use of models. From the surface excess concentration, the area per molecule at interface is calculated using Equation (2.2).

$$A_{min} = \frac{10^{16}}{N\Gamma_{max}} \qquad [2.2]$$

Where, N is Avogadro's number.

The surface tension values at CMC were used to calculate values of surface pressure (effectiveness). The effectiveness of surface tension reduction, $\pi_{CMC} = \gamma_o - \gamma_{CMC}$, where γ_o is the surface tension of water and γ_{CMC} is the surface tension of solution at CMC was determined at different temperatures [167, 168]. The values of π_{CMC} show that, the most efficient one is that which gives the greater lowering in surface tension at the critical micelle concentration. The effectiveness increases with increasing the length of carbon chain in the hydrophobic moiety. The values obtained of (CMC) for nonionic surfactants at different temperatures are given in Table (2.6, 2.7), together with values for the surface tension at CMC (γ_{cmc}). It was found that increasing the number of EO unit decreases CMC. This behavior is based on coiling of the poly (ethylene oxide) chains [166]. In the present system, it was found that the CMC values show an increase with decreasing molecular weight of PEG in the molecule. This can be attributed to the hydrophobic interaction between maleate and oxypropylene groups which increases coiling of poly (oxyethylene) located at the end of the molecules. So the solubility of the surfactants in water is controlled by the structure of hydrophobic groups. It is of interest to mention that the CMC for the prepared surfactant decreases with increasing temperature. This may be attributed to the increase in the radius of gyration of the molecule as a result of increasing the temperature [166]. The same results can be obtained from measuring the cloud temperatures of the prepared surfactants in water. It was established [166] that aqueous solutions of polyoxyethylenated nonionics having oxyethylene content below about 80 wt% become turbid on being heated at a temperature known as the cloud point, above which there is a separation of the solution into two phases. This phase separation occurs in a narrow temperature range (fairly constant) for surfactant concentrations below a few weight percent. The phase appears to consist of an almost micelle-free dilute solution of the nonionic surfactant at a concentration equal to its CMC at this temperature and a surfactant-rich micelle phase, which appears only when the solution is above its cloud point; the two phases merge to form once again a clear solution on cooling. The temperature at which clouding occurs depends on the structure of the polyoxyethylenated nonionic surfactant. The cloud temperatures were measured and are listed in Tables (2.4, 2.5).

Careful inspection of data indicates that, some A_{min} values at the surface decreased with increasing the temperature and this is due to increased dehydration of the hydrophilic group at higher temperature. The effectiveness of adsorption, however, may increase, decrease or show no change with increase in the length of the hydrophobic group depending on the orientation of the surfactant at interface .If surfactant is perpendicular to the surface in a close-packed arrangement, an increase in the length of the straight-chain hydrophobic group appears to cause no significant change in the number of moles of surfactant adsorbed per unit area of surface at surface saturation [169]. This is because the cross-sectional area occupied by the chain oriented perpendicular to the interface does not change with increase in the number of units in the chain. When the area of hydrophilic group is greater than that of the hydrophobic chain, the larger the hydrophilic group, the smaller the amount adsorbed at surface saturation. If the arrangement is predominantly perpendicular but not close-packed, there may be some increase in the effectiveness of adsorption with increase length of hydrophobic group, resulting from greater Van der Waals attraction and consequent closer packing of longer chains [170]. However, if the orientation of surfactant is parallel to the interface, the hydrophobic chain interacts strongly with the surface. e.g., electron rich aromatic nuclei, the effectiveness of adsorption may decrease with increase in the chain length due to increase the cross-sectional area of the molecule on the surface. Thus saturation of the surface will be accomplished by a smaller number of molecules [171]. Finally we can concluded that, the increasing in length of hydrophobic saturated alkyl chain increases the surface excess of molecule and consequently, decreases A_{min} of molecule at air/water interface. This behavior can be attributed to increment of hydrophobic interaction at interface, which increase with increasing length of hydrophobic moieties, which reflects on increasing of surfactant concentration and consequently decreases area per molecule. It is evident that, the minimum area per molecule at air/ water interface can contribute to the molecular area. The surface properties of the prepared PEG-PPO-PEG block copolymers were calculated and listed in Tables (2.6) and (2.7), respectively.

Careful inspection of data, indicates that, A_{min} of the surfactants have two opposite relations with the temperature. The A_{min} may be increased or decreased with increasing the temperature. In polyoxyethylenated nonionics the lack of significant temperature effect may be resulted from two compensating effects [172- 177]:

- Decrease in A_{min} at the surface due to increased dehydration of the hydrophilic group at higher temperature; and
- Increase in A_{min} as a result of enhanced molecular motion at higher temperature.

In the present system, it was found that the minimum area per molecule also increases with increase in temperature, as would be expected from the increased thermal agitation of the molecules in the surface film.

Table (2.6): Surface properties of the prepared PEG-PPO-PEG block copolymers (group 1) at different temperatures

designation	Temp. (K)	cmc Mol/L	γ_{cmc} mN/m	Π_{cmc} mN/m	Γ_{max} x 10^{10} mol/ cm^2	A_{min} nm^2/molecule
PEG400-PPO1-PEG400	298	0.00022	30.8	41.3	**1.70**	0.098
	308	0.00019	29.5	41.6	1.57	0.106
	318	0.00015	29.1	41.0	1.42	0.117
	328	0.00011	28.2	40.9	1.34	0.124
PEG600-PPO1-PEG600	298	0.00014	30	42.1	**1.31**	0.127
	308	0.00035	29	42.1	1.43	0.116
	318	0.00028	28.2	41.9	1.44	0.115
	328	0.00017	27.1	42.0	1.41	0.118
PEG600-PPO2-PEG600	298	0.00014	34.2	37.9	**1.03**	0.161
	308	0.00012	33.5	37.6	0.98	0.169
	318	0.00009	32.2	37.9	0.89	0.187
	328	0.00007	30.1	39.0	0.82	0.202
PEG1500-PPO2-PEG1500	298	0.00008	35.3	36.8	**1.64**	0.101
	308	0.00006	34.5	36.6	1.59	0.104
	318	0.00004	33.3	36.2	1.68	0.099
	328	0.00003	32.2	36.9	1.68	0.099
PEG3000-PPO2-PEG3000	298	0.00006	36	36.1	**1.55**	0.107
	308	0.00004	35	36.1	1.70	0.098
	318	0.00003	34	36.1	1.77	0.094
	328	0.00001	33	36.1	1.79	0.093

Table (2.7): Surface properties of the prepared PEG-PPO-PEG block copolymers (group 2) at different temperatures

Designation	Temp. (K)	cmc x 10^{-6} mMol/L	γ_{cmc} mN/m	Π_{cmc} mN/m	Γ_{max} x 10^{10} mol/ cm^2	A_{min} nm^2/ molecule
PPO-MA(5)-PEG400	298	4.363	36.6	35.5	**1.57**	0.106
	308	2.182	35	36.1	1.59	0.104
	318	1.091	34	36.1	1.58	0.105
	328	1.091	33	36.1	1.53	0.109
PPO-MA(5)-PEG600	298	3.23	44.2	27.9	**1.29**	0.129
	308	1.62	43.3	27.8	1.27	0.131
	318	1.617	42.2	27.9	1.24	0.135
	328	1.62	41.1	28	1.21	0.137
PPO-MA(10)-PEG400	298	4.36	44	28.1	**1.30**	0.127
	308	4.36	42.9	28.2	1.25	0.133
	318	4.36	40.8	29.3	1.26	0.132
	328	4.36	39.9	29.2	1.22	0.136
PPO-MA(10)-PEG600	298	3.23	46.4	25.7	**1.37**	0.122
	308	1.617	45.3	25.8	1.39	0.120
	318	1.617	44.4	25.7	1.33	0.125
	328	3.234	43.5	25.6	1.32	0.126
PPO-MA(20)-PEG400	298	4.36	45	27.1	**1.35**	0.123
	308	4.36	44	27.1	1.32	0.126
	318	2.18	43	27.1	1.28	0.130
	328	2.18	42	27.1	1.23	0.135
PPO-MA(20)-PEG600	298	3.23	48.1	24	**1.54**	0.108
	308	1.617	47	24.1	1.53	0.108
	318	3.23	46	24.1	1.51	0.110
	328	1.617	45	24.1	1.45	0.115

2.5. Thermodynamic Parameters of the Prepared PEG-PPO-PEG Surfactants

The formation of micelles in aqueous solutions is generally viewed as a compromise between the tendency for alkyl chains to avoid energetically unfavorable contacts with water and the desire for the polar parts to maintain contact with the aqueous environment. There are two principally different models for micelle structure. A mean density model [178] is the most appropriate one for micelles consisting of a large core and a relatively thin corona, and star model [179] is the most appropriate for those having a small core from which long chains protrude to form a large corona. The ability for micellization processes depends on thermodynamic parameters, (enthalpy ΔH, entropy ΔS and free energy ΔG) of micellization. Thermodynamic parameters of micillization of the prepared PEG-PPO-PEG block copolymers were calculated and listed in Tables (2.8) and (2.9), respectively. The thermodynamic parameters of micellization are the standard free energies ΔG_{mic}, enthalpies ΔH_{mic}, and entropies ΔS_{mic}, of micellization for nonionic surfactants.

$$\Delta G_{mic} = RT \ln CMC \qquad [2.3]$$

Values of ΔS_{mic} were obtained from Equation (2.4) by invoking the values of ΔG_{mic} at 298, 308, 318 and 328K.

$$\frac{\partial \Delta G_{mic}}{\partial T} = -\Delta S_{mic} \qquad [2.4]$$

In addition, ΔH_{mic}, was calculated from ΔG_{mic} and ΔS_{mic} by applying

$$\Delta H_{mic} = \Delta G_{mic} + T\Delta S_{mic} \qquad [2.5]$$

Table (2.8): Thermodynamic parameters of micellization for the prepared PEG-PPO-PEG block copolymers (group 1)

Surfactants	Thermodynamic Parameters at Different Temperatures								ΔS_{mic}
	298 K		308 K		318 K		328 K		
	$-\Delta G_{mic}$	ΔH_{mic}	$-\Delta G_{mic}$	ΔH_{mic}	$-\Delta G_{mic}$	ΔH_{mic}	$-\Delta G_{mic}$	ΔH_{mic}	
PEG400-PPO1-PEG400	20.1	17.75	21.1	18.016	22.4	17.986	23.9	18.9	0.127
PEG600-PPO1-PEG600	15.7	51.35	19.6	49.7	20.8	50.75	22.8	51.0	0.225
PEG600-PPO2-PEG600	21.2	17.94	22.3	18.05	23.7	17.958	25.1	17.9	0.131
PEG1500-PPO2-PEG1500	22.5	25.776	24	25.896	25.8	25.716	27.3	25.8	0.162
PEG3000-PPO2-PEG3000	23.2	43.85	25	44.3	26.5	45.05	30.2	43.6	0.225

Table (2.9): Thermodynamic parameters of micellization for the prepared PEG-PPO-PEG block copolymers (group 2)

Surfactants	Thermodynamic Parameters at Different Temperatures								ΔS_{mic}
	298 K		308 K		318 K		328 K		
	$-\Delta G_{mic}$	ΔH_{mic}	$-\Delta G_{mic}$	ΔH_{mic}	$-\Delta G_{mic}$	ΔH_{mic}	$-\Delta G_{mic}$	ΔH_{mic}	
PPO-MA(5)-PEG400	29.42	37.95	32.12	37.52	34.93	36.97	36.02	38.14	0.226
PPO-MA(5)-PEG600	30.14	16.44	32.86	15.28	33.92	15.78	34.99	16.28	0.156
PPO-MA(10)-PEG400	29.42	-0.012	30.41	-0.012	31.40	-0.012	32.39	-0.013	0.099
PPO-MA(10)-PEG600	30.14	0.169	32.86	-1.53	33.92	-1.58	33.17	0.186	0.102
PPO-MA(20)-PEG400	29.42	21.50	30.41	22.23	33.16	21.18	34.21	21.85	0.171
PPO-MA(20)-PEG600	30.14	11.17	32.86	9.83	32.16	11.9	34.99	10.47	0.139

The thermodynamic parameters values of adsorption, ΔG_{ad}, ΔH_{ad} and ΔS_{ad} were calculated *via* Equations (2.6), (2.7) and (2.8), respectively [180].

$$\Delta G_{ad} = RT \ln CMC - 0.6023\, \Pi_{CMC} A_{min} \qquad [2.6]$$

$$\frac{\partial \Delta G_{ad}}{\partial T} = -\Delta S_{ad} \qquad [2.7]$$

$$\Delta H_{ad} = \Delta G_{ad} + T\Delta S_{ad} \qquad [2.8]$$

Analyzing the thermodynamic parameters of micellization leads to the fact that micellization process is spontaneous ($\Delta G_{mic} < 0$). The data show that ΔG_{mic} values are less negative with increasing the number of methylene group. This indicates that the increase of hydrophobic groups decreases the micellization process. This can be explained on the basis of steric bulk structure leads to steric inhibition of micellization. On the other hand, the data reveal that $-\Delta G_{mic}$ increases with increasing temperature from 298 to 328 K. The data listed in Tables (2.10) and (2.11) show that ΔS_{mic} values are all positive, indicating increased randomness in the system upon transformation of the nonionic surfactant molecules into micelles or increasing freedom of the hydrophobic chain in the nonpolar interior of the micelles compared to aqueous environment. The decrement of positive ΔS_{mic} value with increasing the chain length units in the surfactant molecule has been observed and can be attributed to increase M.Wt. hydrophobic group leads to decreasing the hydrogen bonds between water and PEG, which increase freedom motion of surfactants. The dissolution of the oxyethylene units has been stated to be the major contributing factor to the positive entropy of micellization in polyoxyethylenated nonionics. An alternative explanation is that there is less restriction on the motion of the surfactant molecule when it is in the essentially water-free environment of the micelle than in the aqueous phase. This extends to both the hydrophobic chain, which is in a hydrocarbon-like environment in the interior of the micelle, and the adjacent part of the hydrophilic polyoxyethylene chain, which is freed, on the only partially solvated micelle surface, from some of the restrictions placed upon it by hydrogen bonding to water molecules. This explanation, which assigns the change in entropy to the solute rather than to the solvent, is consistent with the re-evaluation of the concept [181] of entropy of solution.

The values of ΔG_{ad}, ΔH_{ad} and ΔS_{ad} for both group 1 and group 2 nonionic surfactants are calculated and listed in Tables (2.10) and (2.11), respectively. All ΔG_{ad} values are more negative than ΔG_{mic}, indicating that adsorption at the interface is associated with a decrease in the free energy of the system. This may be attributed to the effect of steric factor on inhibition of micellization more than its effect on adsorption. The values of ΔS_{ad} are all positive and have some values greater than ΔS_{mic} for nonionic surfactants in group 1 but ΔS_{ad} have greater values than ΔS_{mic} for nonionic surfactants in group 2 . This may reflect the greater freedom of motion of the hydrophobic chains at the planar air- aqueous solution interface compared to that in the relatively cramped interior beneath of the convex surface of the micelle. This indicates that the steric factor inhibits micellization more than do adsorption for nonionic surfactants. On the other hand, almost all positive values of ΔH_{mic} are much greater than the corresponding values of ΔH_{ad}. This indicates that the dehydration-breaking of hydrogen bonds- at adsorption is easier than at micellization. The negative values of ΔH_{ad} indicate that more bonds between polyoxyethylene chain oxygen and water molecules are broken in the process of adsorption at the air aqueous solution interface than in micellization.

Table (2.10): Thermodynamic parameters of adsorption for the prepared PEG-PPO-PEG block copolymers (group 1)

Surfactants	Thermodynamic Parameters at Different Temperatures								ΔS_{ad}
	298 K		308 K		318 K		328 K		
	$-\Delta G_{ad}$	ΔH_{ad}	$-\Delta G_{ad}$	ΔH_{ad}	$-\Delta G_{ad}$	ΔH_{ad}	$-\Delta G_{ad}$	ΔH_{ad}	
PEG400-PPO1-PEG400	22.5	21.8	23.8	22.01	25.3	21.98	27	21.8	0.15
PEG600-PPO1-PEG600	18.9	45.45	22.5	44.03	23.7	44.99	25.7	45.15	0.216
PEG600-PPO2-PEG600	24.8	25.6	26.1	26.02	28	25.8	29.8	25.6	0.169
PEG1500-PPO2-PEG1500	24.7	23.3	26.2	23.4	28	23.3	29.5	23.3	0.162
PEG3000-PPO2-PEG3000	25.5	38.57	27.1	39.12	28.5	39.87	32.2	38.8	0.215

Table (2.11): Thermodynamic parameters of adsorption for the prepared PEG-PPO-PEG block copolymers (group 2)

Surfactants	Thermodynamic Parameters at Different Temperatures								ΔS_{ad}
	298 K		308 K		318 K		328 K		
	$-\Delta G_{ad}$	ΔH_{ad}	$-\Delta G_{ad}$	ΔH_{ad}	$-\Delta G_{ad}$	ΔH_{ad}	$-\Delta G_{ad}$	ΔH_{ad}	
PPG-MA(5)-PEG400	31.69	36.52	34.39	36.11	37.21	35.58	38.38	36.70	0.229
PPG-MA(5)-PEG600	32.30	15.89	35.04	14.76	36.18	15.24	37.31	15.73	0.162
PPG-MA(10)-PEG400	31.58	0.217	32.66	0.201	33.73	0.197	34.78	0.217	0.107
PPG-MA(10)-PEG600	32.02	-0.999	34.72	-2.65	35.85	-2.75	35.11	-0.966	0.104
PPG-MA(20)-PEG400	31.43	21.53	32.47	22.26	35.28	21.23	36.42	21.87	0.178
PPG-MA(20)-PEG600	31.70	10.68	34.43	9.37	33.76	11.46	36.66	9.98	0.142

2.6. Evaluations of the Prepared Surfactants as Demulsifiers for Different Types of Synthetic Crude Oil Emulsions

The co-production of water and crude oil in the form of an emulsion is highly undesirable, such emulsions introduce technical challenges so they must be resolved to provide the specified product quality [182]. Crude oil is always produced as a persistent water-in-oil emulsion, which may be resolved into two separate phases before the crude can be accepted for pipelining. The water droplets are sterically stabilized by the asphaltene and resin fractions of the crude oil. These are condensed aromatic rings containing saturated carbon chains and naphthenic rings as substituents, along with a distribution of heteroatoms and metals. They are capable of crosslinking at the water drop-oil interface. Thus, a stable emulsion exists only when emulsifying agents are present. Elimination, alteration, or neutralization of the emulsifying agents will allow immiscible liquids to separate. Addition of suiTable chemicals with demulsifying properties specific to the crude oil to be treated will generally provide quick, cost-effective, and flexible resolution of emulsions. Crude oil emulsions are stabilized by high-molecular weight surfactants, viz., asphaltenes and resins. They do not develop high surface pressures, and therefore steric stabilization of water-in-crude-oil emulsions is the most plausible mechanism of stabilization of such emulsions. Demulsifier molecules and natural surfactants compete with each other for adsorption onto the water-drop film. When demulsifier molecules, which lower interfacial tension much more than the natural surfactants, are adsorbed at the interface, the film becomes unstable in the direction of coalescence of water drops. The addition of various demulsifiers produced several effects on the destabilization of water in crude oil emulsions. The efficiency of a surfactant to act as a demulsifier depends on many factors are the distribution of the surfactant throughout the bulk volume of emulsion, the partitioning of the surfactant between the phases, the temperature, pH, and salt content of aqueous phase. Other factors of importance are the mode of injection of the surfactant, the concentration of the surfactant, type of solvent carrier, amount of water in emulsion and age of emulsion. In this study one has also taken into account that the type of oil plays a role in the demulsification process. The oil samples have been sampled from different production wells to be used as representative sample for wells. In this respect, many studies have been carried on stability and demulsification of crude oil emulsions reported by many investigators [183-186]. Demulsification by use of chemicals or demulsifiers is a very complex phenomenon. Demulsifiers displace the natural stabilizers (emulsifiers) present in the interfacial film around the water droplets. This displacement is brought about by the adsorption of the demulsifier at the interface. This displacement occurring at the oil / water interface, influence the coalescence of water droplets through enhanced film drainage. The efficiency of the demulsifier is dependent on its adsorption at the oil / water, or droplet surface [187].

The present work investigates the efficiency of the prepared surfactants as demulsifier. It has been shown that in emulsions any change in temperature causes changes in the interfacial tension between two phases (oil and water) [188], in the nature and viscosity of the interfacial film [189], in the relative solubility of the emulsifying agent in the two phases [190], and in the thermal agitation of the dispersed droplets [191]. Therefore, temperature changes usually cause considerable changes in the stability of the emulsion; they may invert the emulsion or cause it to break up [192]. The effect of the salinity of the aqueous phase on the stability of oil-in-water type emulsions has been extensively studied by several authors. The activity of demulsifiers was compared with respect to their state before addition to the emulsion (dissolved or neat). The study revealed that a demulsifier, which is dissolved composition and chemical structure of the prepared surfactants on the in a solvent, gives a better separation of the phases than does an undiluted demulsifier. In the present work, we investigate the effect of the solvent, temperature, concentration of the surfactant, crude oil/water emulsion demulsification efficiencies [193-195].

2.6.1. Demulsification performance of prepared PEG-*PPO-PEG* block copolymers

Previous work [196] studied the effect of solvent on the demulsification efficiencies of crude oil emulsions. This study revealed that, the xylene and alcohol mixtures possess better demulsification efficiencies. The role of solvent on the demulsification efficiency of crude oil emulsion is of primary importance and has been briefly discussed by Canevari. It was found that the demulsification performance of a demulsifier is based on the interaction between dissoluted surfactants and water droplet through diffusion and adsorption, which permits faster transport of demulsifier molecules to the water droplet interface. Moreover, dissoluted demulsifiers give a better separation of the phases (water and crude oil phases) than an undiluted demulsifier does [197]. In the present work, we investigate the effect of the solvent used to dissolute the prepared surfactants with respect to its ability to break up water-in-oil emulsions. Accordingly, xylene/ethanol mixture (1:1) is selected as solvent for the present surfactants. This mixture is the best solvent and can be explained in terms of being much more efficient in solubilizing polar and nonpolar moieties in the demulsifier molecule. The polar moiety represented by PEG units is stabilized by ethanol, while the nonpolar moiety (hydrocarbon moiety) is solubilized by xylene. This enhanced solubilization is reflected in a better dispersion in crude oil-water emulsion and consequently in a better demulsification performance. The testing results were obtained from bottle test. The testing procedures were carried out at four different concentrations of the used demulsifiers (50, 100, 250 and 500 ppm) at 333 K and at 318 K for some surfactants on four synthetic water-in-crude-oil emulsions which were pronounced at different ratios of crude oil: water (90:10, 80:20, 70:30 and 50:50) respectively. The dehydration rate of each surfactant was tested at four different concentrations (50, 100, 250 and 500 ppm). The five prepared surfactants in group 1 showed good efficiency as dehydrating agents for water in crude oil emulsions. Also the six prepared surfactants in group 2 showed good efficiency as dehydrating agents for water in crude oil emulsions. The testing results showed in some cases that there was no constant concentration for all surfactants at which the best dehydration was given; some surfactants reached the best dehydration rate at 500 ppm and the others at 250 ppm or even 100 ppm.

2.6.1.1. Effect of demulsifiers on demulsification efficiency for the four prepared crude oil emulsions

A comparison between the best dehydration rates for each surfactant at a given concentration was plotted for the four types of the prepared emulsions. It is observed that the five surfactants in group 1 have the same dehydration speed which was in the order PEG600-PPO1-PEG600 > PEG400-PPO1-PEG400> PEG600-PPO2-PEG600 > PEG1500-PPO2-PEG1500> PEG3000-PPO2-PEG3000, where PEG600-PPO1-PEG600 and PEG400-PPO1-PEG400 gave the same dehydration rate 100% but at different time at all. **Figure (2.13 a)** shows the dehydration rate of the five surfactants prepared in group 1 at 333 K for O: W 90:10. It was observed that the maximum dehydration rate was 100% at concentration 500 ppm, from the plotted data it was shown that PEG600-PPO2-PEG600 gave 70% but the dehydration rate of PEG1500-PPO2-PEG1500 and PEG3000-PPO2-PEG3000 stopped at 64% and 58% respectively. This behavior can correlated to good solubility of PEG600-PPO1-PEG600 and PEG400-PPO1-PEG400 surfactants in oil continuous phase (90:10) of emulsion which reflects on good adsorption at oil water interface. **Figure (2.13 b)** shows the difference in dehydration rate between the five surfactants at 333 K for O: W 80:20. It can be observed that the dehydration rate of PEG600-PPO2-PEG600 is 67% while the dehydration rate of PEG1500-PPO2-PEG1500 and PEG3000-PPO2-PEG3000reached 60% and 53% respectively.

Figure (2.13 c) shows the dehydration rate of the five surfactants at 333 K for O:W 70:30. From the plotted data it was noticed that three surfactants PEG600-PPO2-PEG600, PEG1500-PPO2-PEG1500 and PEG3000-PPO2-PEG3000 reached the dehydration rate 65%, 54% and 46% respectively. This can be refereed to increasing of surfactant water solubility in above order which increase with decreasing the length of hydrophobic $(CH)_2$ chain. This can be elucidated from increasing water content in W/O emulsion to be 50: 50. **Figure (2.13 d)** shows the dehydration rate of the five surfactants at 333 K for O: W 50:50. It was noticed that, the activity of the prepared surfactants was reduced when the water content riches 30% and 50 %. The solubility of non-ionic surfactants in water increases with increasing their HLB values. It is noticed that the HLB values of the prepared surfactants were 5.1-13. This indicates the partial solubility of most of these surfactants in water. The low HLB values of the prepared surfactants will be more soluble in non-polar solvents [197]. The prepared emulsions can be considered as water in oil emulsions, in other words the oil percent in these emulsions is more than water since the prepared surfactants are more soluble in oil than water, so it will be expected to give good dehydration rates with such emulsions. Due to the little difference in HLB values of the prepared surfactants it is difficult to discuss the effect of HLB completely in comparing the dehydration rates of the five surfactants with all types of W/O emulsions. Finally, we found that the efficiency decreased by increasing water percent from O: W 90: 10, 80: 20 and 70: 30 to O: W 50:50. These results agree with the data reported on adsorption parameters and IFT measurements **Figure (2.14)** which indicate that PEG600-PPO1-PEG600 surfactant has best adsorption performance and high reduction in IFT.

In group 2 surfactants, **Figure (2.15 a)** shows the dehydration rate of the six surfactants at 333 K for O: W 90:10. It was observed that the maximum dehydration rate was 100%, the dehydration speed was in the order: PPO-MA(5)-PEG400 > PPO-MA(10)-PEG400 > PPO-MA(5)-PEG600 > PPO-MA(10)-PEG600 > PPO-MA(20)-PEG400 > PPO-MA(20)-PEG600. From the plotted data it was shown that PPO-MA(5)-PEG400 gave dehydration rate 100% and PPO-MA(10)-PEG400 gave 70% where the dehydration rate PPO-MA(5)-PEG600 was 66% but PPO-MA(10)-PEG600 and PPO-MA(20)-PEG400 have the same dehydration rate at 58% and PPO-MA(1)-PEG600 at 45%. **Figure (2.15 b)** shows the difference in dehydration rate between the six surfactants at 333 K for O: W 80:20. It was noticed that the dehydration speed was in the order: PPO-MA(5)-PEG400 > PPO-MA(10)-PEG400 > PPO-MA(20)-PEG400 > PPO-MA(5)-PEG600 > PPO-MA(10)-PEG600 > PPO-MA(20)-PEGME600. It can be observed that the maximum dehydration rate reached 100% with PPO-MA(5)-PEG400. While the dehydration rate of , PPO-MA(10)-PEG400 , PPO-MA(20)-PEG400, PPO-MA(5)-PEG600, PPO-MA(10)-PEG600 and PPO-MA(20)-PEGME600 reached 72%, 61%, 60%, 59% and 49% respectively at concentration 250 ppm. **Figure (2.15 c)** shows the dehydration rate of the six surfactants at 333 K for O: W 70:30. It was noticed that the dehydration speed was in the order: PPO-MA(5)-PEG400 > PPO-MA(10)-PEG400 > PPO-MA(20)-PEG400 > PPO-MA(5)-PEG600 > PPO-MA(10)-PEG600 > PPO-MA(20)-PEGME600. From the plotted data it was noticed that the maximum dehydration rate reached 100% with PPO-MA(5)-PEG400. While the dehydration rate of , PPO-MA(10)-PEG400 , PPO-MA(20)-PEG400, PPO-MA(5)-PEG600, PPO-MA(10)-PEG600 and PPO-MA(20)-PEGME600 reached 74%, 63%, 61%, 58% and 51% respectively at concentration 500 ppm. **Figure (2.15 d)** shows the dehydration rate of the six surfactants at 333 K for O: W 50:50. It was noticed that, the dehydration rate reached 100% with PPO-MA(5)-PEG400, PPO-MA(10)-PEG400 and PPO-MA(20)-PEG400 while the dehydration rate of PPO-MA(5)-PEG600 was 98% and PPO-MA(10)-PEG600 and PPO-MA(20)-PEGME600 have the same dehydration rate which was 96% . This can be refereed to increasing of surfactant water solubility in above order which increase with decreasing the length of hydrophobic $(CH)_2$ chain. This can be elucidated from increasing water content in W/O emulsion. Surfactants in group 2 gave the best dehydration rate with O:W 50:50 this may be also due to the effect of HLB on its solubility in water more than the other surfactants, so it showed the best dehydration rate with the emulsion that contains the highest percent of water. But as mentioned before due to the little difference in HLB values of the five surfactants the effect of HLB on their efficiency is not completely clear.

Tables (2.12 & 2.13) show the demulsification rate of different crude oil emulsion using different concentrations of the prepared PEG-PPO-PEG surfactants at 333 K for group 1 and group 2, respectively.

Table (2.12): demulsification rate of different crude oil emulsions using different concentrations of PEG-PPO-PEG surfactants (group 1) at 333 K

Surfactants	Oil/water emulsion 90 /10 at 333 K			Blank
	500 ppm	250 ppm	100 ppm	
	Water separated ml% / time, minute			
PEG400-PPO1-PEG400	100/50	100/300	64/300	0
PEG600-PPO2-PEG600	70/300	60/300	50/300	
PEG1500-PPO2-PEG1500	64/300	46/300	42/300	
PEG3000-PPO2-PEG3000	58/300	30/300	24/300	
PEG600-PPO1-PEG600	100/45	100/60	100/140	
	Oil/water emulsion 80/20 at 333 K			
PEG400-PPO1-PEG400	100/60	100/130	82/300	0
PEG600-PPO2-PEG600	67/300	54/300	49/300	
PEG1500-PPO2-PEG1500	60/300	44/300	40/300	

PEG3000-PPO2-PEG3000	53/300	28/300	23/300	
PEG600-PPO1-PEG600	100/50	100/100	85/300	
	Oil/water emulsion 70/30 at 333 K			
PEG400-PPO1-PEG400	100/120	100/160	73.3/300	0
PEG600-PPO2-PEG600	65.3/300	57.3/300	46.7/300	
PEG1500-PPO2-PEG1500	54/300	42/300	38.7/300	
PEG3000-PPO2-PEG3000	46/300	26.7/300	22/300	
PEG600-PPO1-PEG600	100/65	100/120	83.3/300	
	Oil/water emulsion 50/50 at 333 K			
PEG400-PPO1-PEG400	100/150	100/300	72/300	0
PEG600-PPO2-PEG600	62.8/300	56.8/300	44.8/300	
PEG1500-PPO2-PEG1500	52/300	40/300	32/300	
PEG3000-PPO2-PEG3000	45.6/300	25.6/300	21.2/300	
PEG600-PPO1-PEG600	100/100	100/300	80/300	

Table (2.13): demulsification rate of different crude oil emulsions using different concentrations of PEG-PPO-PEG surfactants (group 2) at 333 K

Surfactants	Oil/water emulsion 90/10 at 333 K			Blank
	500 ppm	250 ppm	100 ppm	
	Water separated ml% / time, minute			
PPO-MA(5)-PEG400	100/120	75/180	70/180	0
PPO-MA(5)-PEG600	66/180	50/180	39/180	
PPO-MA(10)-PEG400	70/180	58/180	42/180	
PPO-MA(10)-PEG600	58/180	40/180	30/180	

PPO-MA(20)-PEG400	58/180	40/180	30/180	
PPO-MA(20)-PEG600	45/180	23/180	20/180	
	80/20 at 333 K			
PPO-MA(5)-PEG400	60/180	100/110	65/180	0
PPO-MA(5)-PEG600	37/180	60/180	41/180	
PPO-MA(10)-PEG400	39/180	72/180	41/180	
PPO-MA(10)-PEG600	35/180	59/180	39/180	
PPO-MA(20)-PEG400	36/180	61/180	40/180	
PPO-MA(20)-PEG600	21/180	49/180	23/180	
	70/30 at 333 K			
PPO-MA(5)-PEG400	100/90	59/180	40/180	0
PPO-MA(5)-PEG600	61/180	52/180	37/180	
PPO-MA(10)-PEG400	74/180	54/180	37/180	
PPO-MA(10)-PEG600	58/180	49/180	35/180	
PPO-MA(20)-PEG400	63/180	54/180	39/180	
PPO-MA(20)-PEG600	51/180	44/180	23/180	
	50/50 at 333 K			
PPO-MA(5)-PEG400	100/120	100/105	93/180	0
PPO-MA(5)-PEG600	92/180	98/180	84/180	
PPO-MA(10)-PEG400	100/145	100/120	92/180	
PPO-MA(10)-PEG600	94/180	96/180	82/180	
PPO-MA(20)-PEG400	100/155	100/135	88/180	
PPO-MA(20)-PEG600	88/180	96/180	80/180	

The dehydration rate of group 1 and group 2 surfactants was tested at 318 K for O:W 90:10 to study the effect of temperature on demulsification efficiency. **Figure**

(**2.16**) shows that the dehydration speed of the two surfactants is in the order PEG600-PPO1-PEG600 > PEG400-PPO1-PEG400, the dehydration rate of PEG600-PPO1-PEG600 reached 78%, while that of PEG400-PPO1-PEG400 was 50%, while **Figure (2.17)** shows that the dehydration speed of the two surfactants is in the order PPO-MA(5)-PEG400 > PPO-MA(10)-PEG400, the dehydration rate of reached PPO-MA(5)-PEG400 66%, at 180 minute while that of PPO-MA(10)-PEG400 was 43% at the same dehydration time. This behavior reflects the effect of cloud points on the demulsification efficiency. In general **Figures (2.16 & 2.17)** showed that there are steady states at which the dehydration rate has constant values for some time, and then it started to increase again till reaching the final percent of dehydration for each surfactant. To understand this behavior, the mechanism of demulsification should be discussed. The proposed mechanism for water separation of crude oil based on PEO-PPO-PEO demulsifiers can be illustrated in **Figure (2.18)**. In this respect, Crude oil emulsions are stabilized by surfactants, viz., asphaltenes and resins. They do not develop high surface pressures, and therefore steric stabilization of water-in-crude-oil emulsions is the most plausible mechanism of stabilization of such emulsions. Demulsifier molecules and natural surfactants compete with each other for adsorption onto the water-drop film.

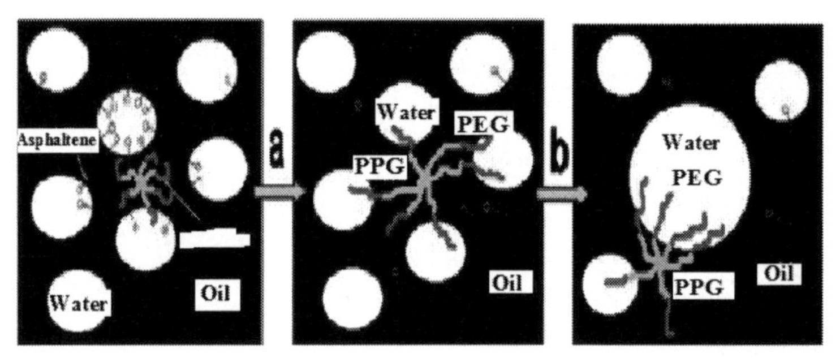

Figure (2.18): Physical model of water drops caught by PEG-PPO-PEG copolymer: (a) displacement and (b) coalescence

When demulsifier molecules, which lower interfacial tension much more than the natural surfactants, are adsorbed at the interface, the film becomes unstable in the direction of coalescence of water drops. Emulsions formed in the petroleum industry are predominantly water-in-oil or regular emulsions, in which the oil is the continuous or external phase and the dispersed water droplets, form the dispersed or internal phase. The water droplets become bigger with time after demulsifier is added.

The role of the demulsifier is to change the interfacial properties and to destabilize the surfactant-stabilized emulsion film in the demulsification process. In the beginning, small and uniform drops flocculate and some large drops begin to form at 30 min. Then two or more large drops continue to form a single larger drop: coalescence happens. The droplet size grows fast and the droplet number reduces after demulsifier is added. It was concluded that coalescence of water droplets destroyed emulsions. Three terms related to stability commonly encountered in crude oil emulsion are "flocculation" "coalescence" and "breaking". Although they are sometimes used almost interchangeably, those terms are in fact quite distinct in meaning as far as the condition of an emulsion is concerned. "Flocculation" refers to the mutual attachment of individual emulsion drops to form flocs or loose assemblies. Flocculation can be in many cases a reversible process overcomes by the input of much less energy than was required in the original emulsification process. "Coalescence" refers to the joining of two or more drops to form a single drop of greater volume, but smaller interfacial area. Although coalescence will result in significant microscopic changes in the condition of the dispersed phase, it may not immediately result in a macroscopically apparent alteration of the system. The "breaking" of an emulsion refers to a process in which gross separation of the two phases occurs. In such an event, the identity of individual drops is lost, along with the physical and chemical properties of the emulsion. This process obviously represents a true loss of stability in the emulsion. It has been established that the kinetics of chemical demulsification is complicated by the interaction of three main effects. These are the displacement of the asphaltenic film from the oil water interface by the demulsifier; flocculation and coalescence of water drops [187]. **Figure (2.19)** compares between the dehydration rates of PEG600-PPO1-PEG600 with the four types of emulsions at 333 K, which has high HLB value 12.1 showed the best dehydration rate with the emulsion that contains low percent of water. But as mentioned before due to the little difference in HLB values of the five surfactants the effect of HLB on their efficiency is not completely clear.

2.6.1.2. Effect of concentration of demulsifiers on demulsification efficiency

The dehydration rate of each surfactant was tested at four different concentrations (50, 100, 250 and 500 ppm). The testing results showed that there was no constant concentration for all surfactants at which the best dehydration was given; some surfactants reached the best dehydration rate at 250 ppm and the others at 500 ppm or even 100 ppm. **Figures (3.20 a-e)**, **(2.21 a-e)**, **(2. 22 a-e)** and **(2.23 a-e)** for group 1 and **(2.24 a-f)**, **(2.25a-f)**, **(2.26 a-f)** & **(2.27 a-f)** for group 2 showed the effect of concentration on demulsification efficiency of each surfactant(demulsifier) on different types of crude oil emulsions of O:W (90:10, 80:20, 70:30 and 50:50), respectively. The plotted data showed a comparison between the used concentrations of each demulsifier on the dehydration rate of that demulsifiers for any type of emulsions. It was observed that there is a specific concentration for each demulsifier at which it gives the best dehydration rate. These concentrations ranges from 50-500 ppm, some demulsifiers gave good results at 250 or 100 ppm while the others gave the best dehydration rates at higher concentrations.

CHAPTER III

"FIGURES"

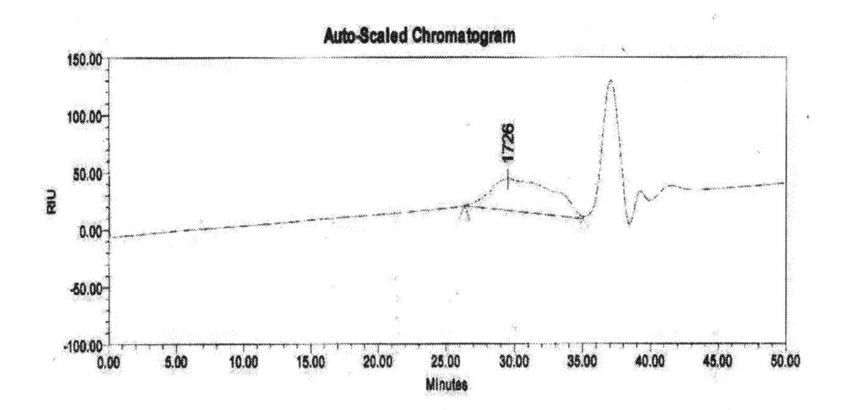

GPC Results

	Retention Time	Mn	Mw	MP	Mz	Mz+1	Polydispersity
1	29.537	416	1348	1726	3089	4783	3.241846

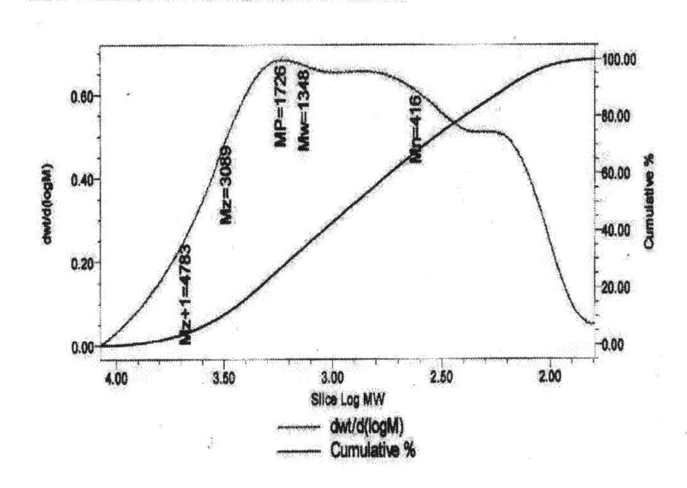

Fig. (2.1): GPC data of PPO1

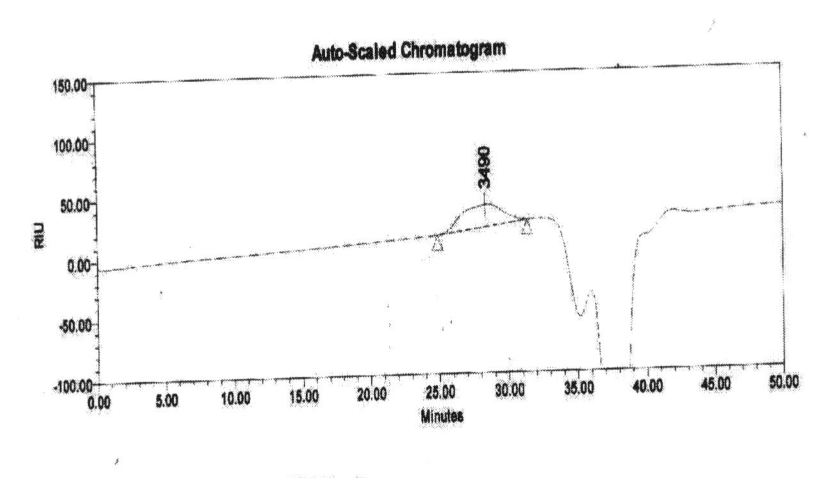

GPC Results

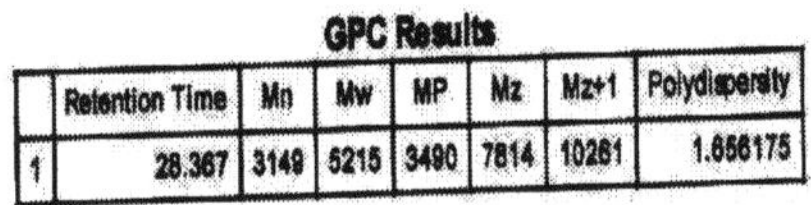

	Retention Time	Mn	Mw	MP	Mz	Mz+1	Polydispersity
1	28.367	3149	5215	3490	7814	10261	1.656175

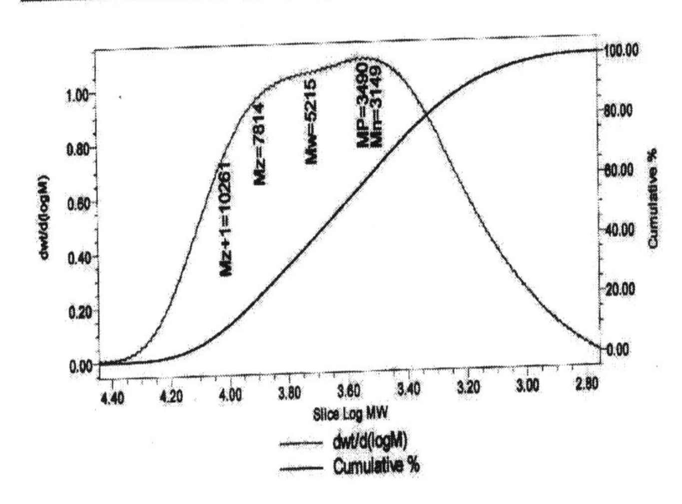

Fig. (2.2): GPC data of PPO2

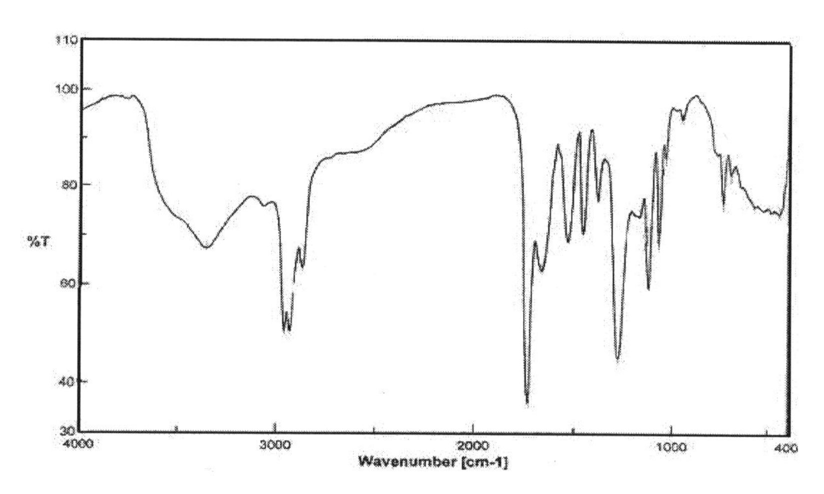

Fig. (2.3): IR spectrum of PEG600-PPO1-PEG600

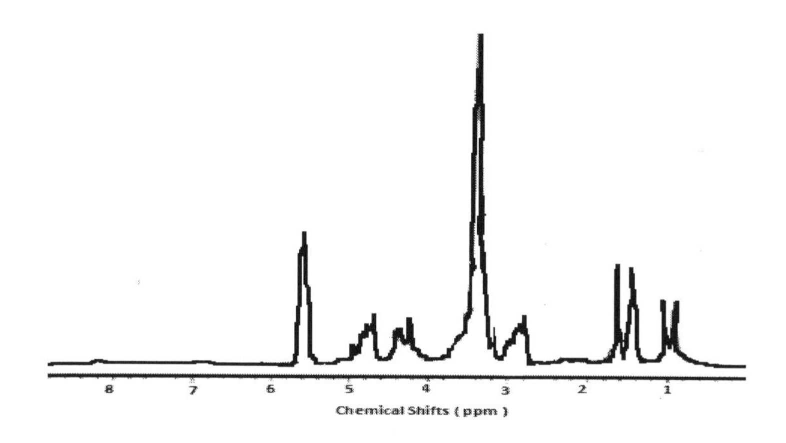

Fig. (2.4): ^{1}H-NMR Spectrum of PEG600 –PPO2-PEG600

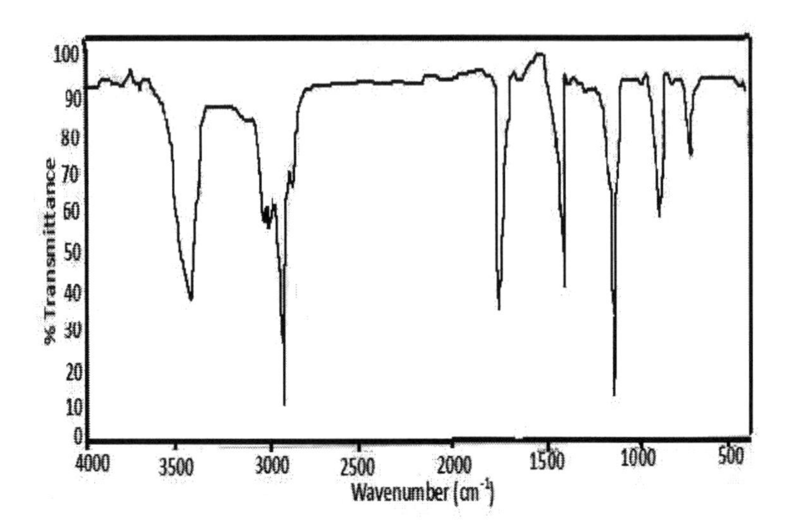

Fig. (2.5.a): IR spectrum of PEG400-PPO1-PEG400

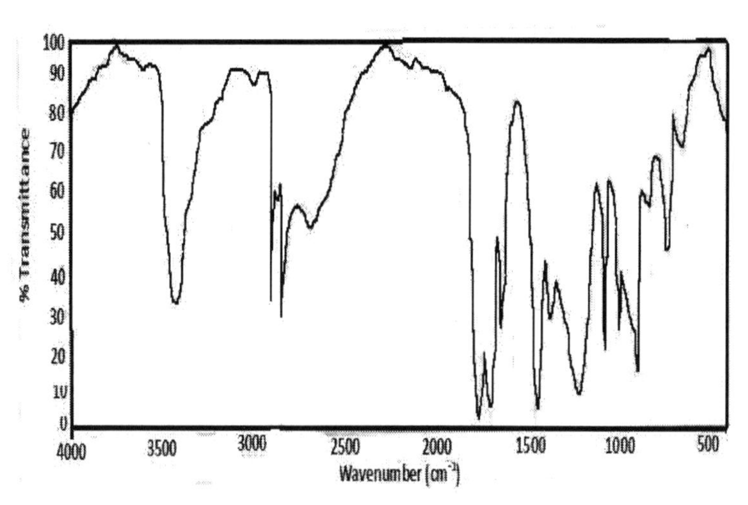

Fig. (2.5.b): IR spectrum of PEG600-PPO2-PEG600

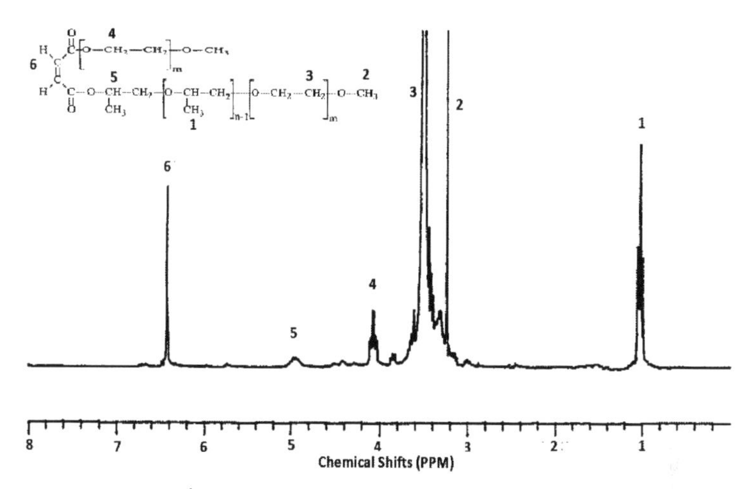

Fig. (2.6.a): ^{1}H-NMR Spectrum of PEG400 –PPO1-PEG400

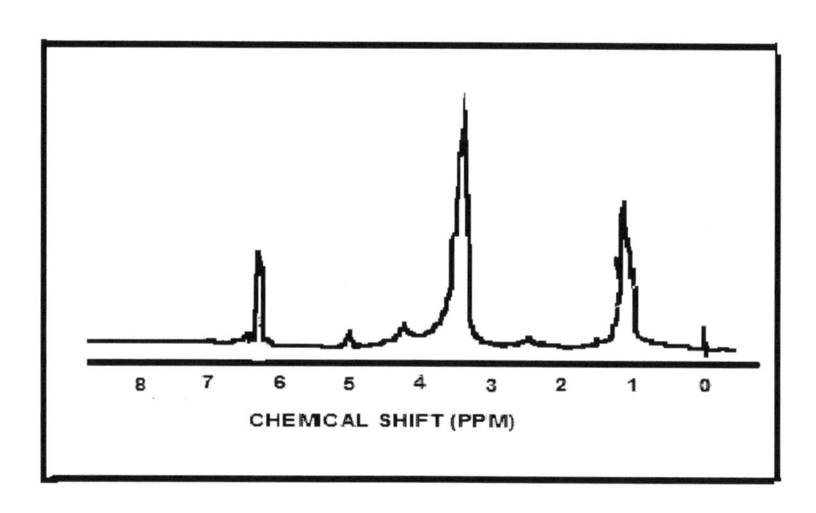

Fig. (2.6.b): ^{1}H-NMR Spectrum of PEG600 –PPO1-PEG600

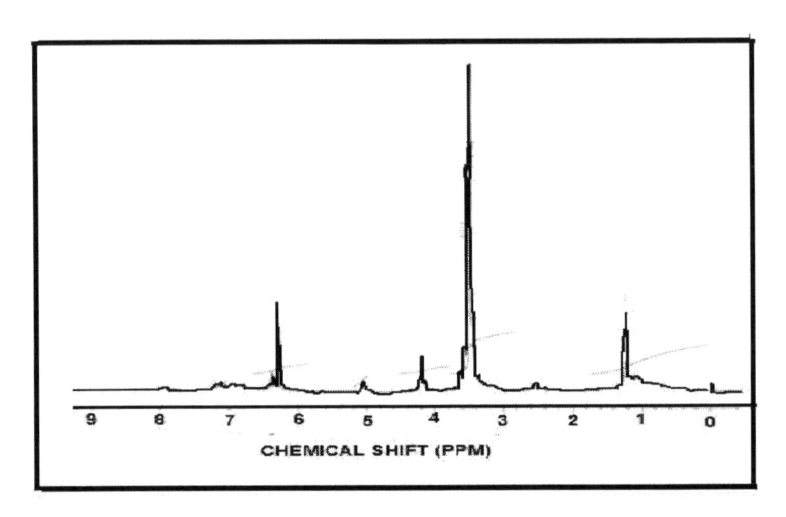

Fig. (2.6.c): ^{1}H-NMR Spectrum of PEG1500 –PPO2-PEG1500

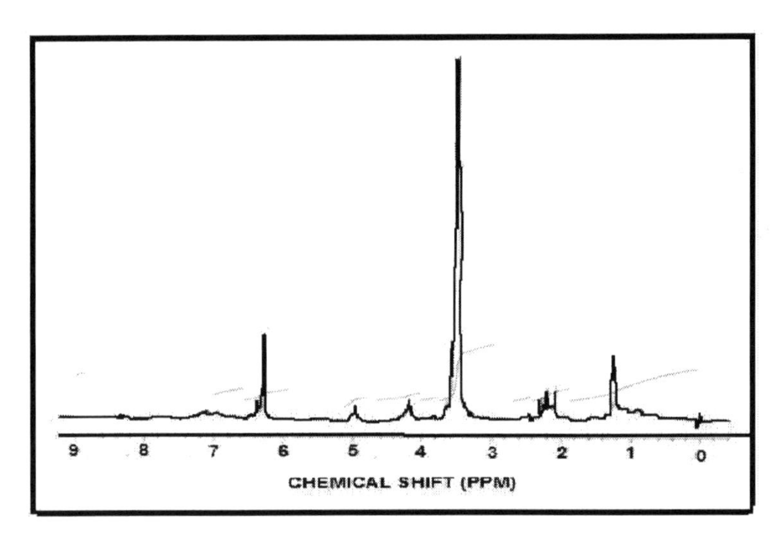

Fig. (2.6.d): ^{1}H-NMR Spectrum of PEG3000 –PPO2-PEG3000

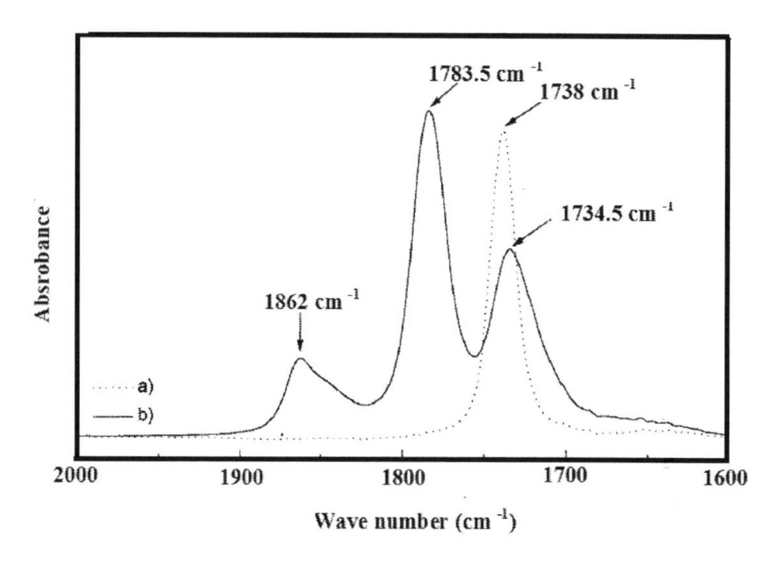

Fig. (2.7): FT-IR Spectra of a) PPO-Ac and b) PPO-MA10

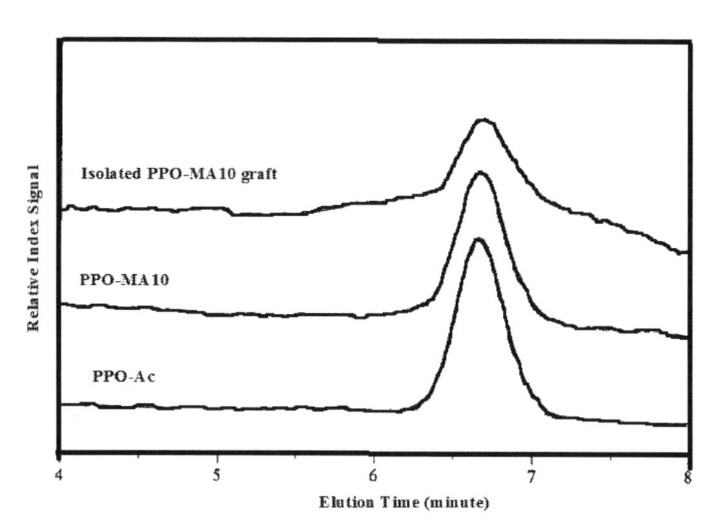

Fig. (2.8): GPC curves (refractive signal curves) of PPO-Ac, PPO-MA10 and isolated highly grafted parts of the reaction products of PPO-MA10

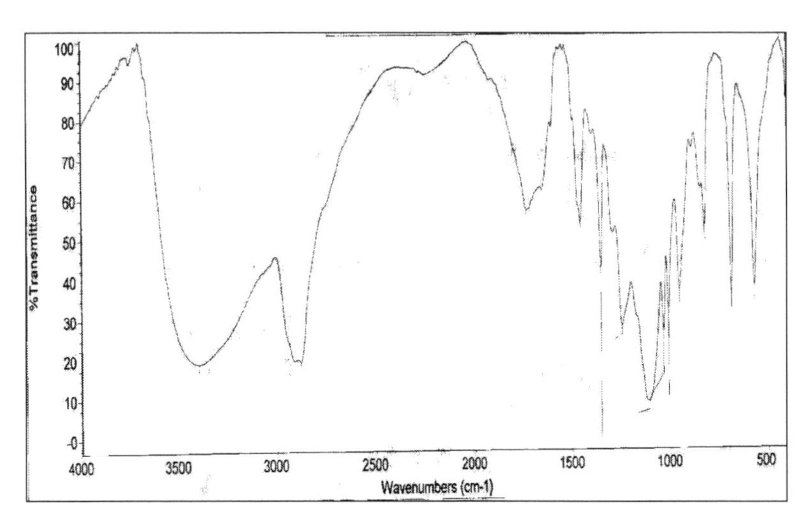

Fig. (2.9.a): FTIR Spectrum of PPO-MA5-PEG600

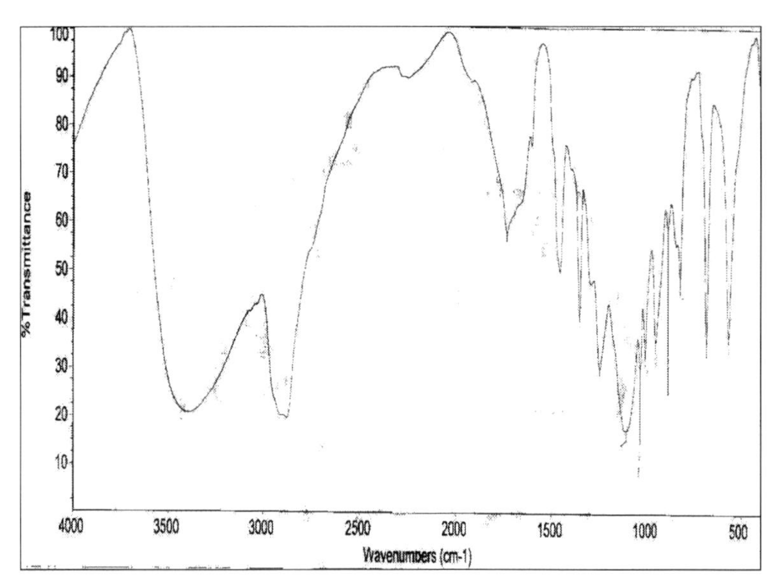

Fig. (2.9.b): FTIR Spectrum of PPO-MA10-PEG400

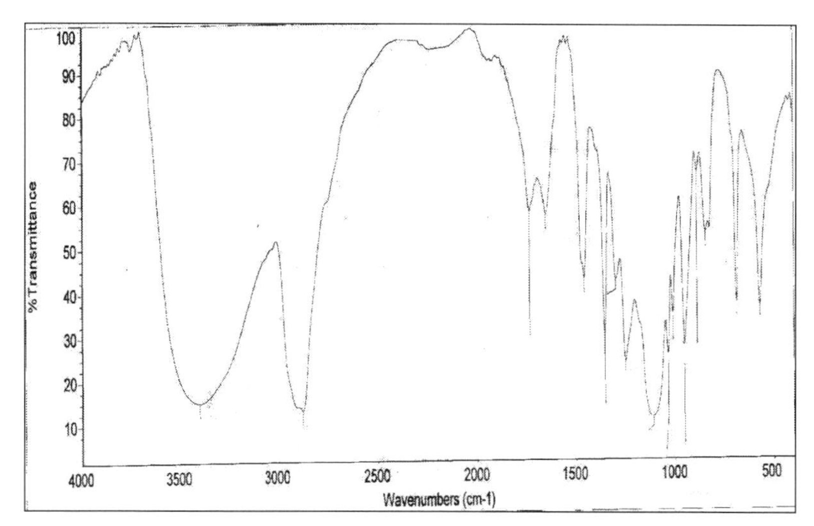

Fig. (2.9.c): FTIR Spectrum of PPO-MA10-PEG600

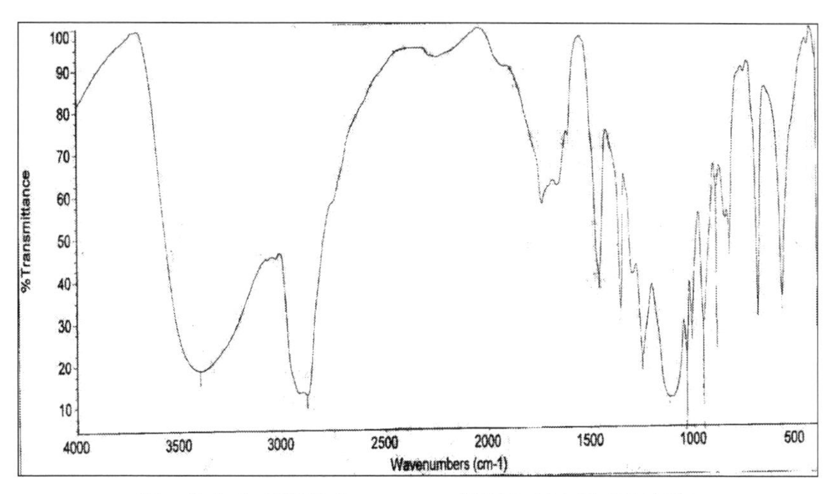

Fig. (2.9.d): FTIR Spectrum of PPO-MA20-PEG400

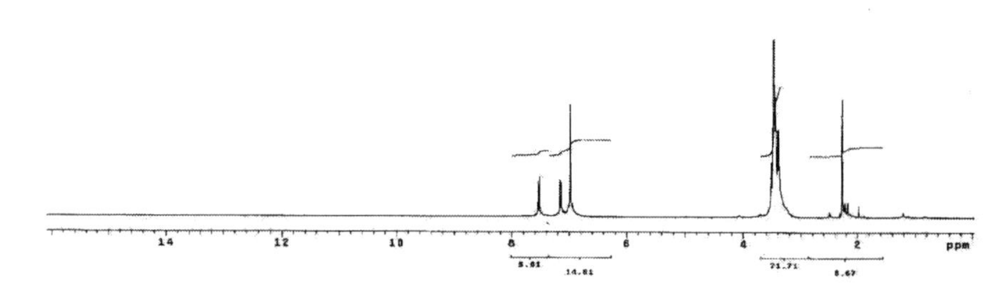

Fig. (2.10.a): ^{1}H-NMR Spectrum of PPO-MA5-PEG400

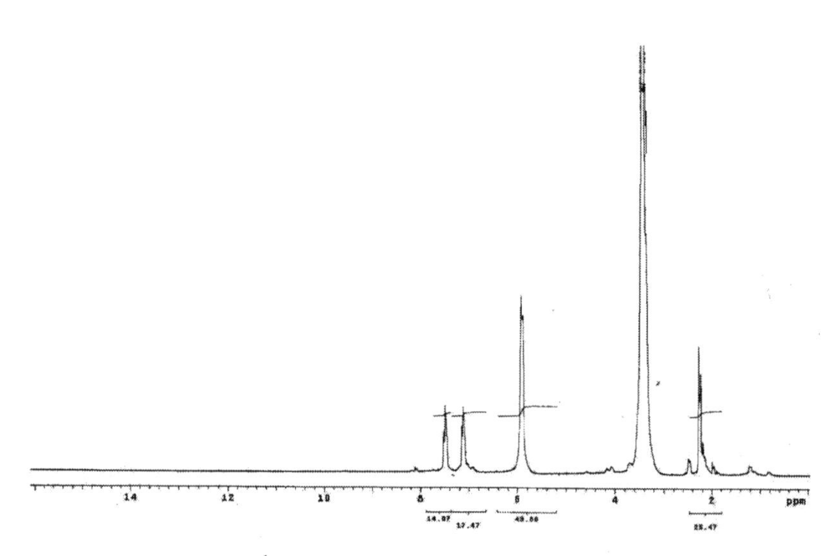

Fig. (2.10.b): ^{1}H-NMR Spectrum of PPO-MA10-PEG400

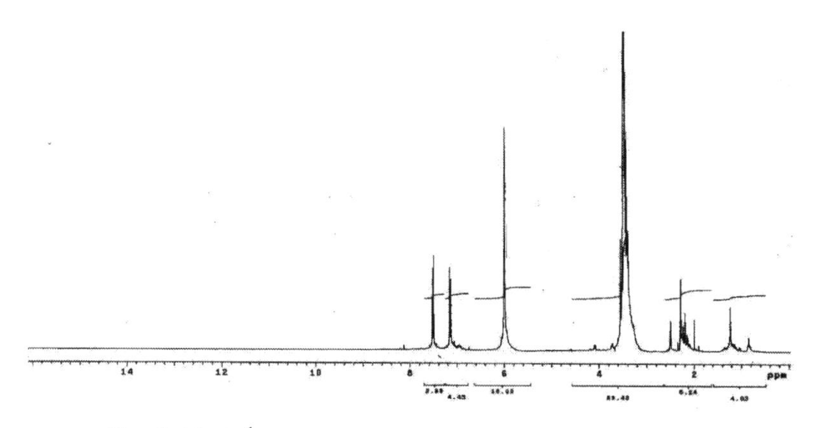

Fig. (2.10.c): ¹H-NMR Spectrum of PPO-MA20-PEG400

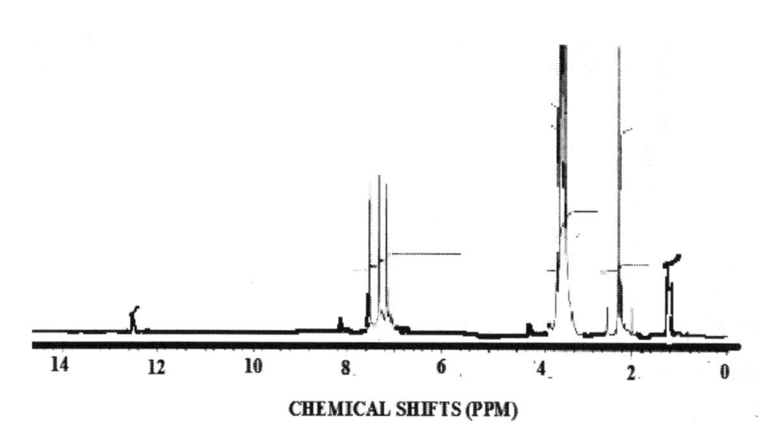

Fig. (2.10.d): ¹H-NMR Spectrum of PPO-MA20-PEG600

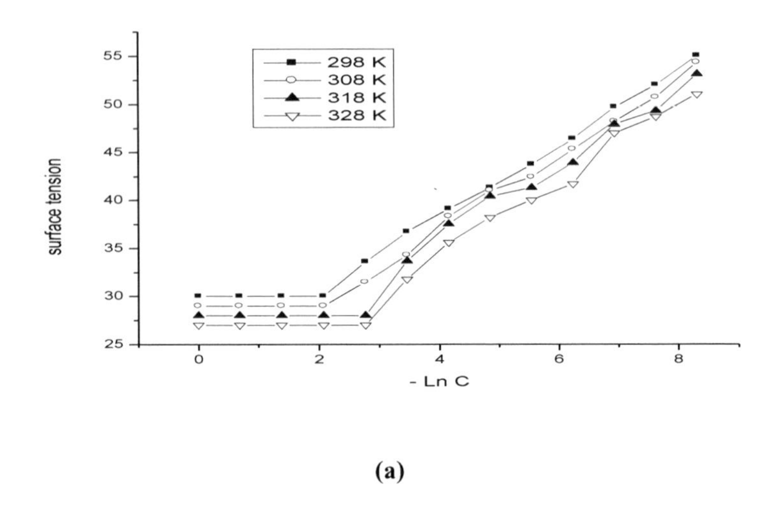

(a)

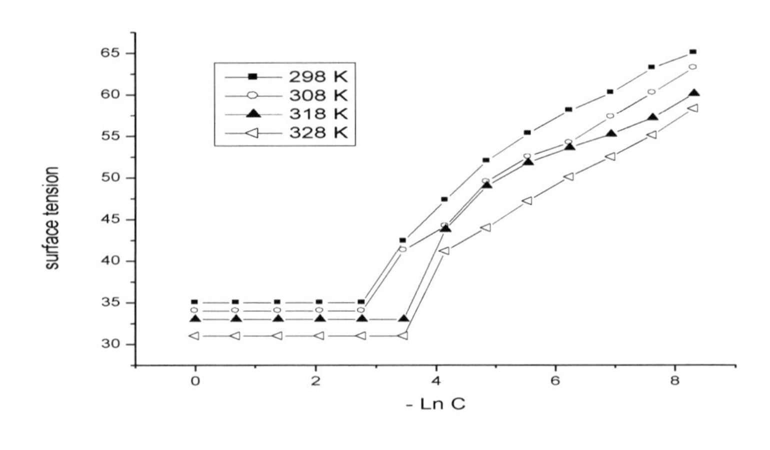

(b)

Fig. (2.11.a,b): Adsorption isotherms of a) PEG400-PPO1-PEG400 and b) PEG600-PPO1-PEG600 at 298, 308, 318 and 328 K

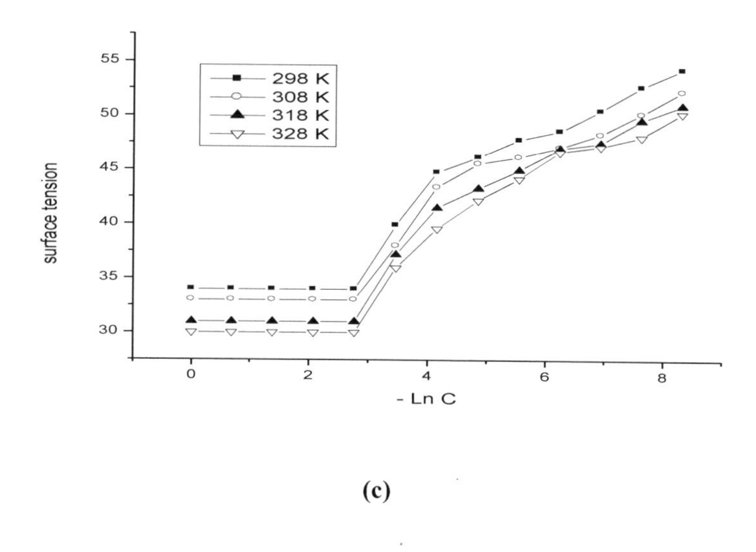

(c)

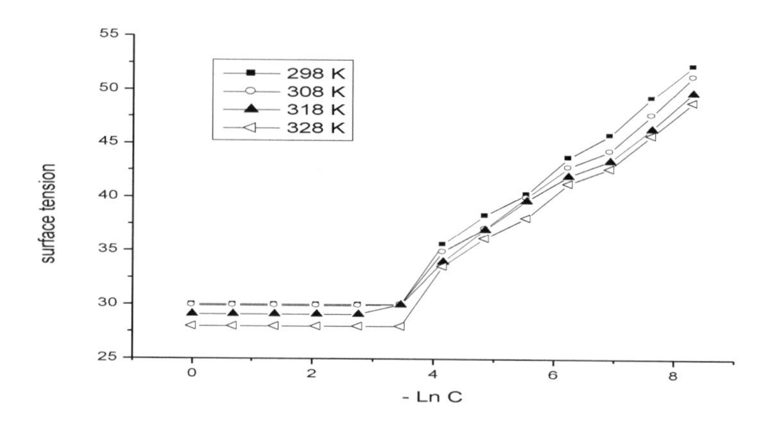

(d)

Fig. (2.11.c,d): Adsorption isotherms of c) PEG600-PPO2-PEG600 and d) PEG1500-PPO2-PEG1500 at 298, 308, 318 and 328 K

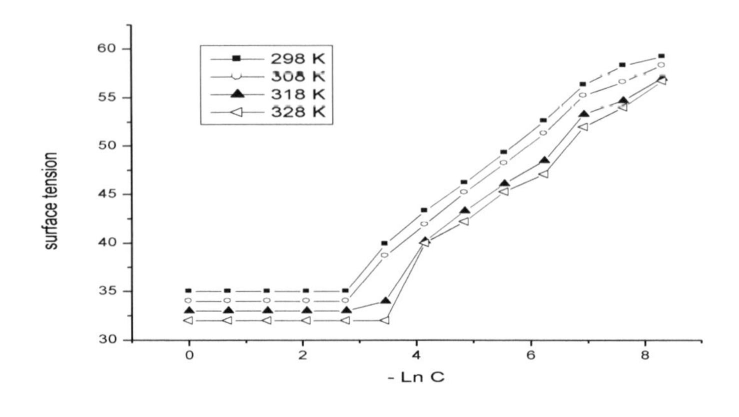

(e)

Fig. (2.11.e): Adsorption isotherms of e) PEG3000-PPO2-PEG3000 at 298, 308, 318 and 328 K.

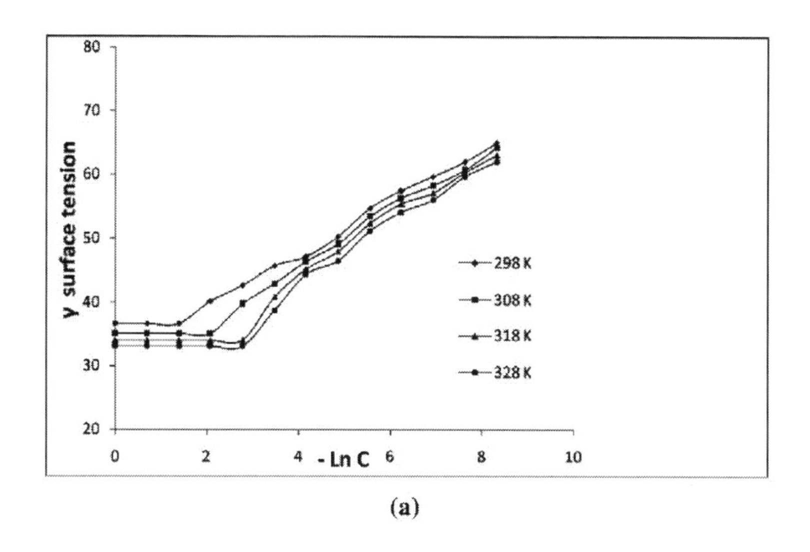

(a)

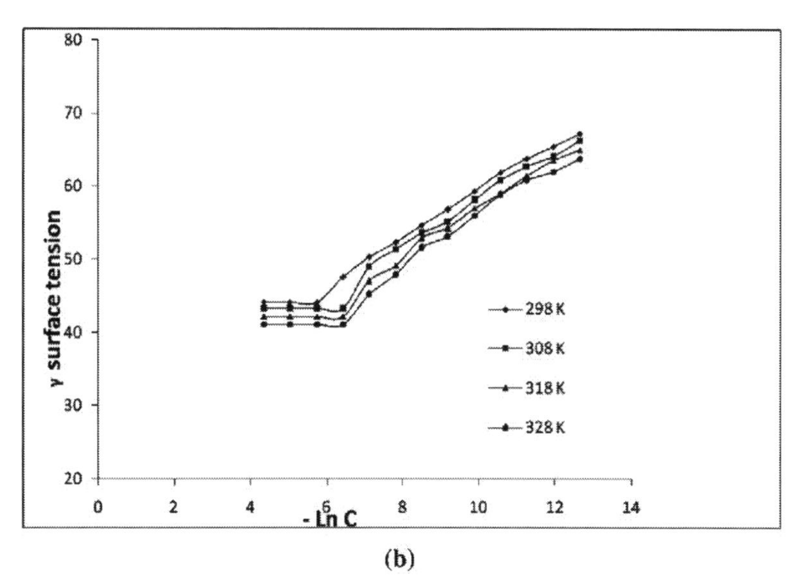

(b)

Fig. (2.12.a, b): Adsorption isotherms of a) PPO2-MA5-PEG400 and b) PPO2-MA5-PEG600 at 298, 308, 318 and 328 K

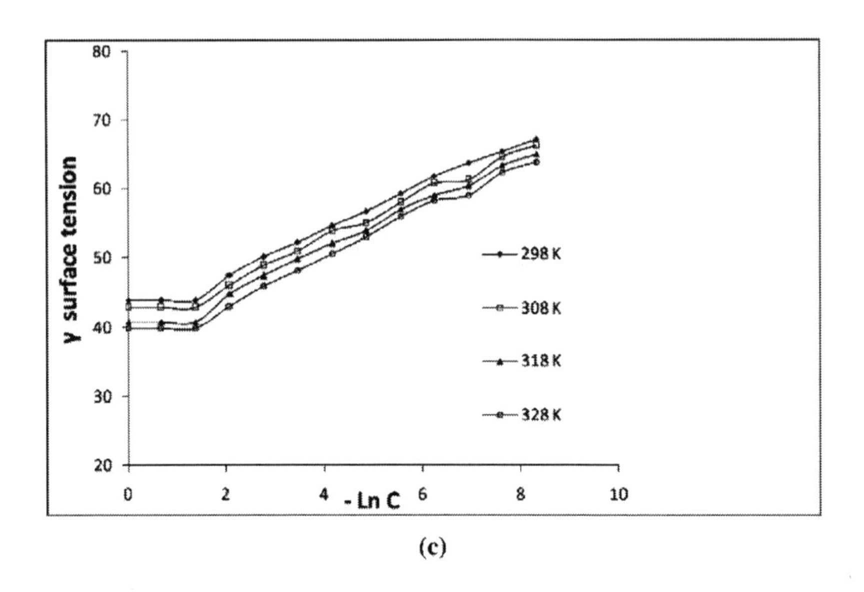

(c)

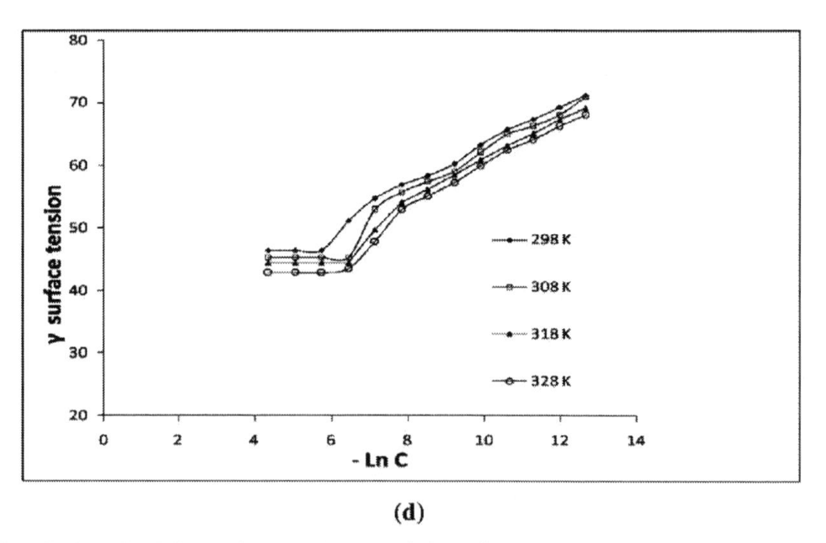

(d)

Fig. (2.12.c, d): Adsorption isotherms of c) PPO2-MA10-PEG400 and d) PPO2-MA10-PEG600 at 298, 308, 318 and 328 K

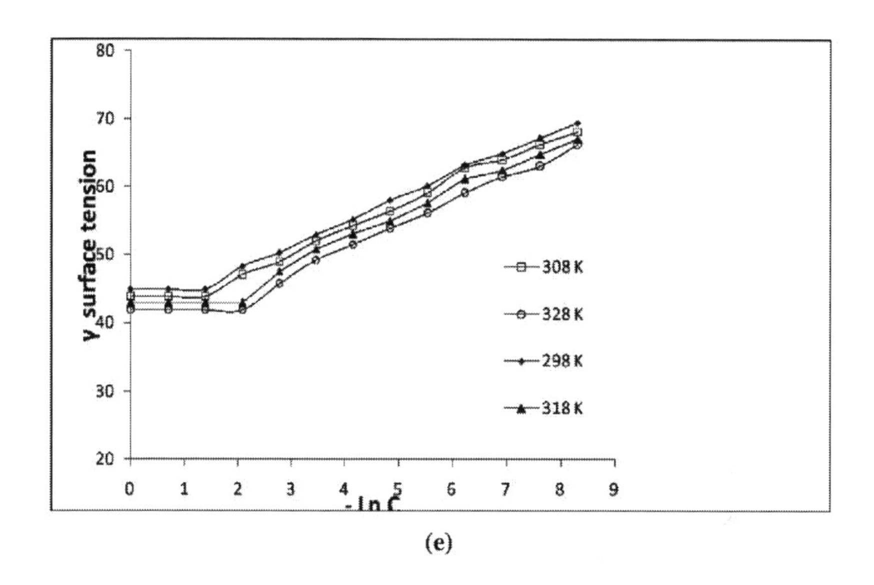

(e)

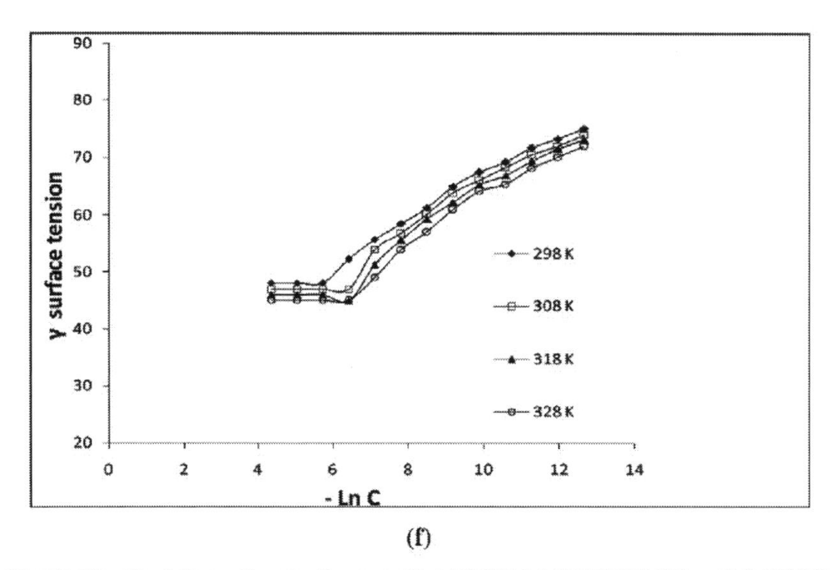

(f)

Fig. (2.12.e,f): Adsorption isotherms of e) PPO2-MA20-PEG400 and f) PPO2-MA20-PEG600 at 298, 308, 318 and 328 K

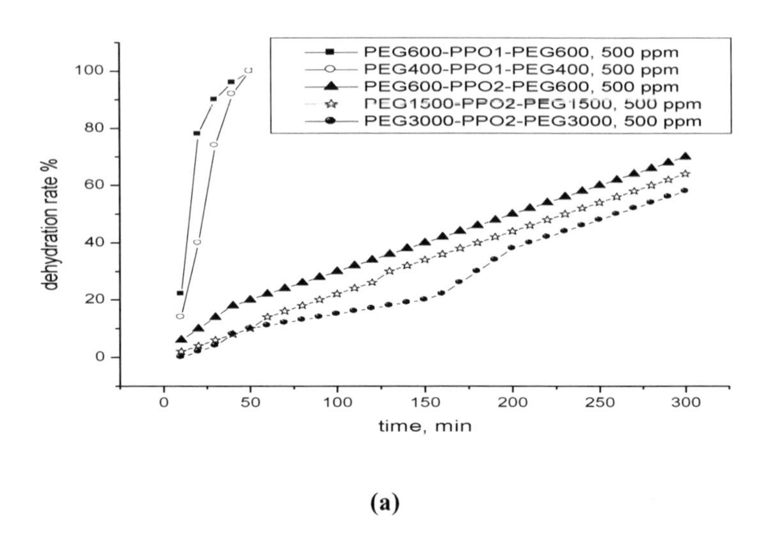

(a)

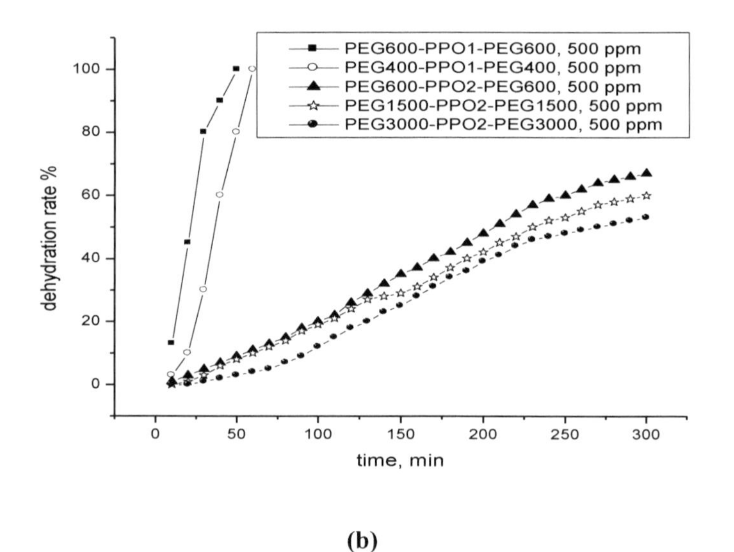

(b)

Fig. (2.13.a, b): Effect of Demulsifiers on Dehydration Rate of a) 90:10 and b) 80:20 Crude Oil: Water Emulsions at 333 K

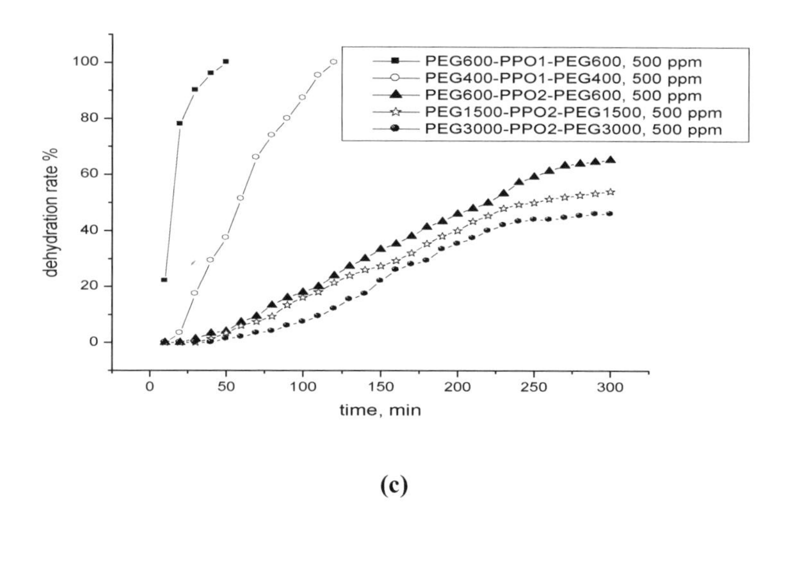

(c)

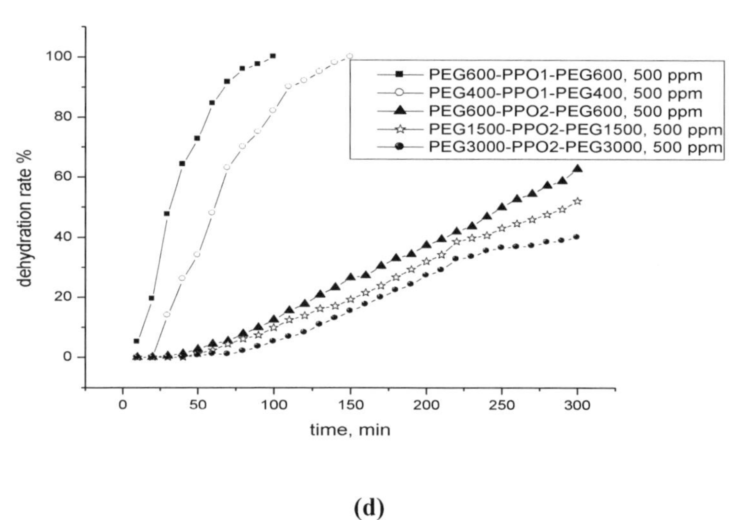

(d)

Fig. (2.13.c, d): Effect of Demulsifiers on Dehydration Rate of c) 70:30 and d) 50:50 Crude Oil: Water Emulsions at 333 K

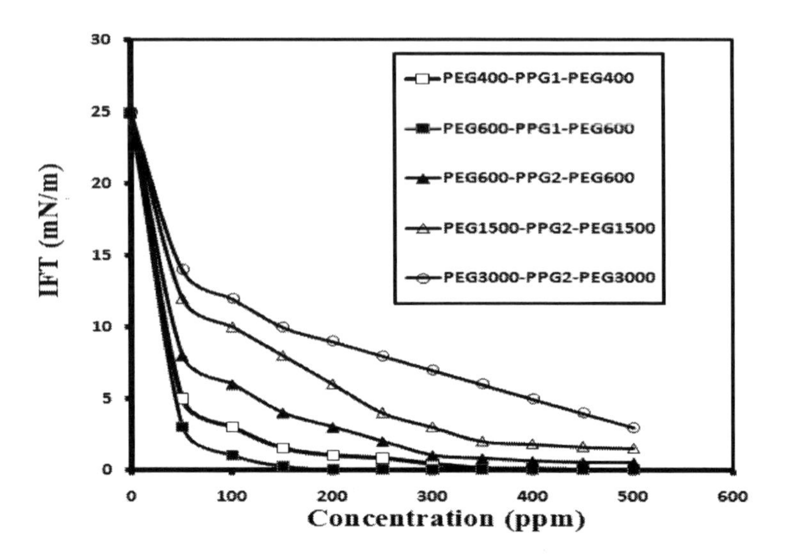

Fig. (2.14): The effect of demulsifier concentrations on interfacial tension of crude oil/water (90/10) interfaces at 298 K

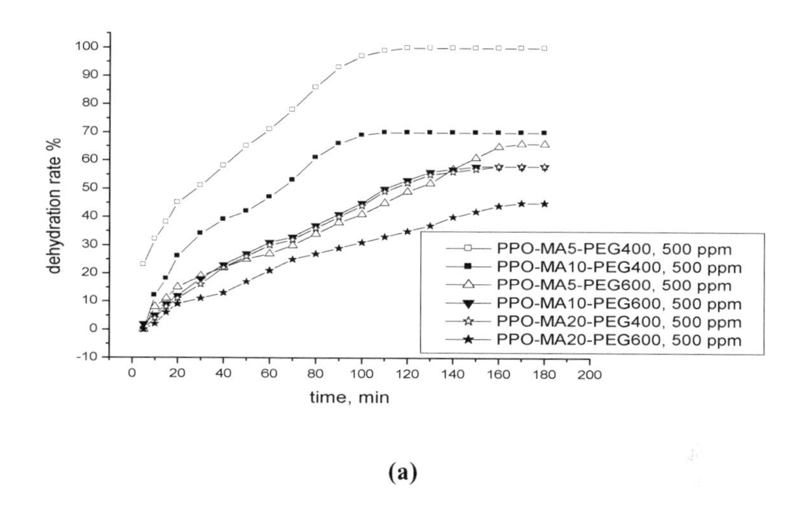

(a)

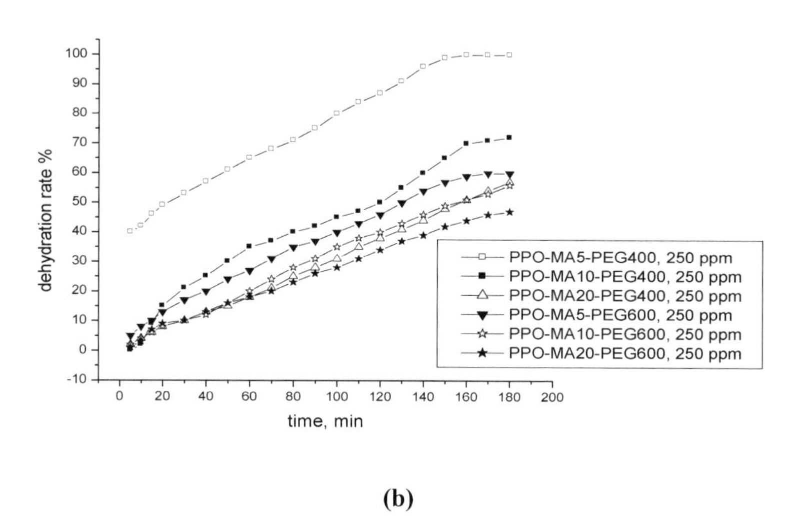

(b)

Fig. (2.15. a,b): Effect of Demulsifiers on Dehydration Rate of a) 90:10 and b) 80:20 Crude Oil: Water Emulsions at 333 k

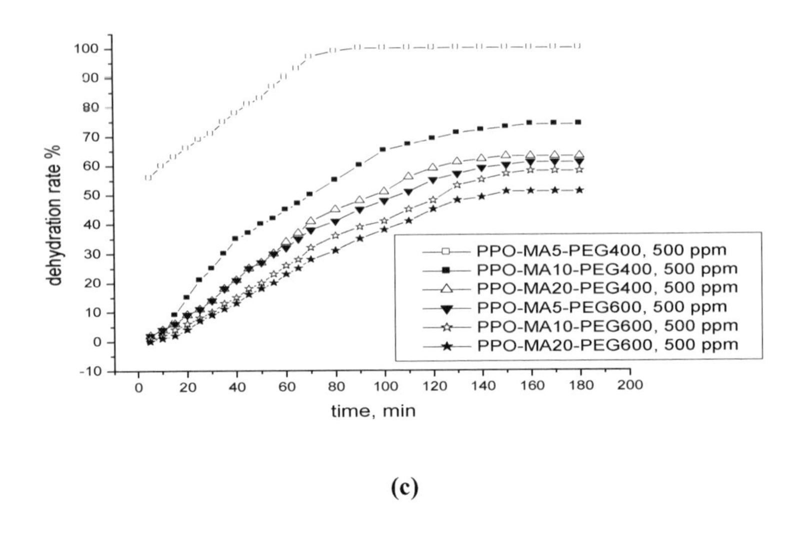

(c)

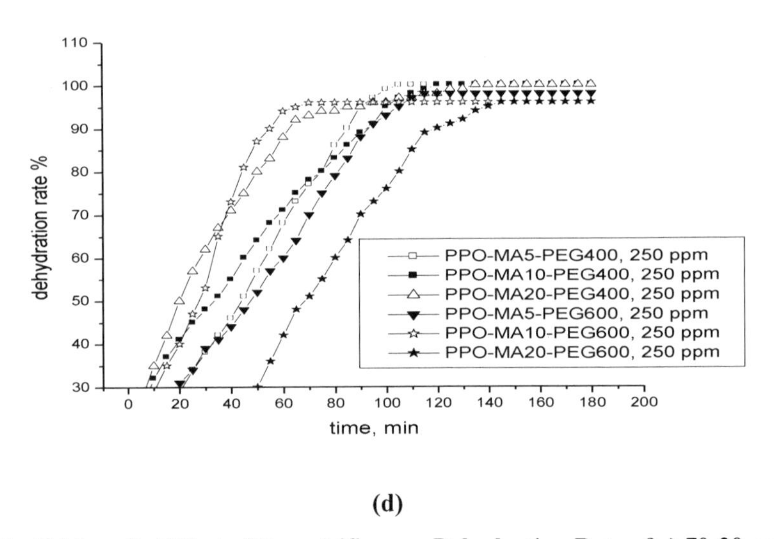

(d)

Fig. (2.15. c, d): Effect of Demulsifiers on Dehydration Rate of c) 70:30 and d)50:50Crude Oil: Water Emulsions at 333 k

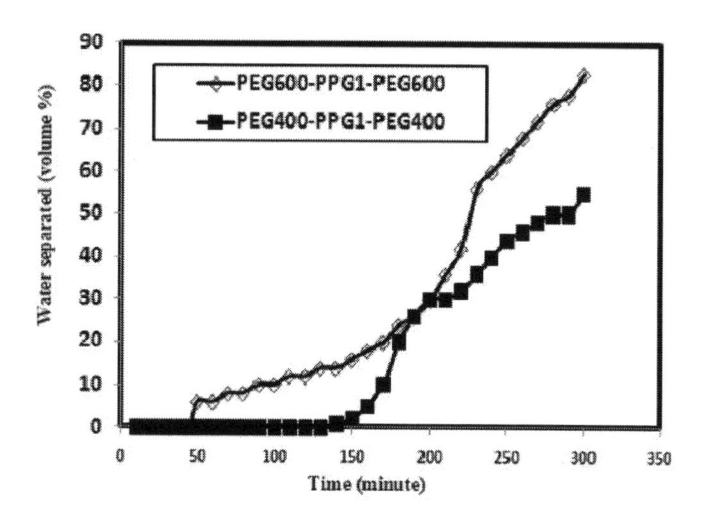

Fig. (2.16): Effect of Demulsifiers (group 1) on Dehydration Rate of 90:10 Crude Oil: Water Emulsions at at 318 k

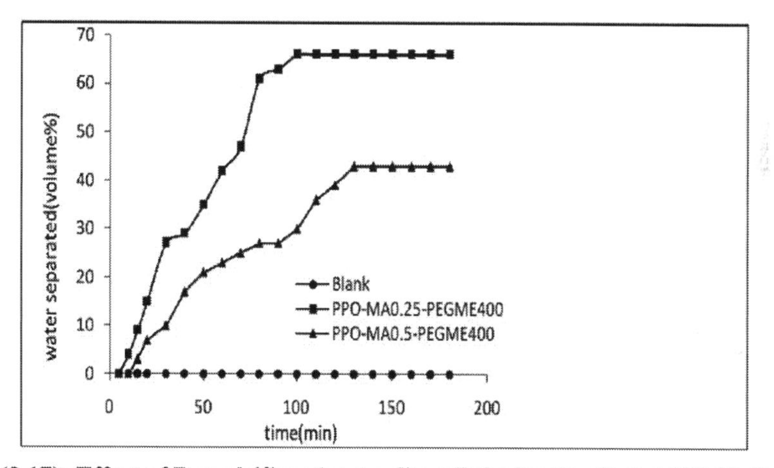

Fig. (2.17): Effect of Demulsifiers (group 2) on Dehydration Rate of 90:10 Crude Oil: Water Emulsions at 318 k

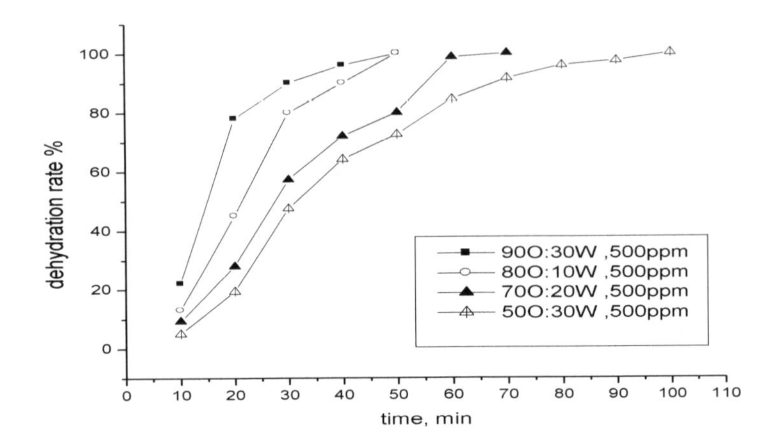

Fig. (2.19): Effect of PEG600-PPO1-PEG600 demulsifier on dehydration rate of different crude oil: water emulsions at 333 k

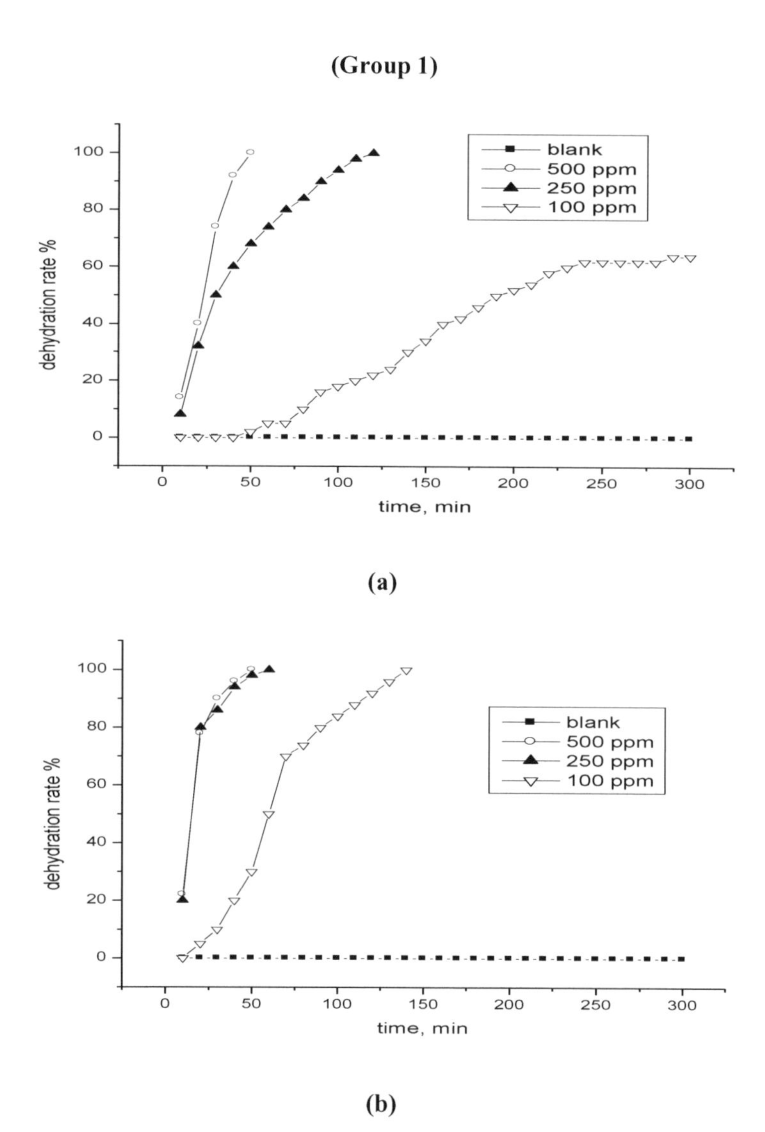

Fig. (2.20.a,b): Effect of Concentration on the Dehydration Rate of a) PEG600-PPO1-PEG600 and b) PEG400-PPO1-PEG400 at 333 K for an O:W Emulsion 90:10

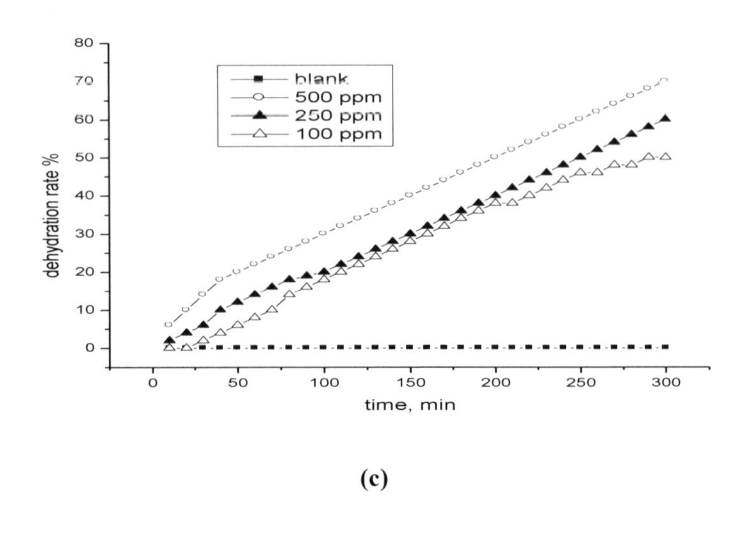

(c)

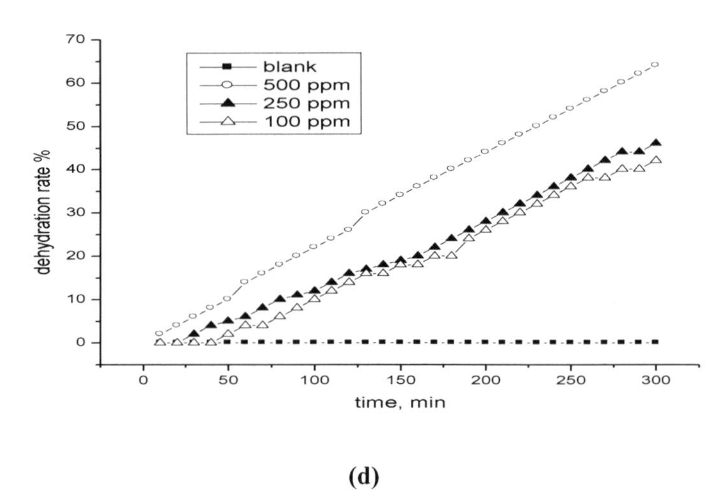

(d)

Fig. (2.20. c,d): Effect of Concentration on the Dehydration Rate of c) PEG600-PPO2-PEG600 and d) PEG1500-PPO2-PEG1500 at 333 K for an O:W Emulsion 90:10.

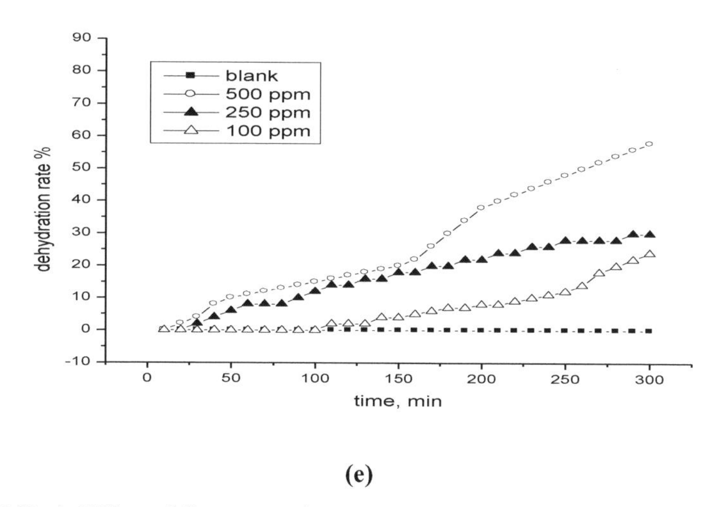

(e)

Fig. (2.20.e): Effect of Concentration on the Dehydration Rate of e) PEG3000-PPO2-PEG3000 at 333 K for an O:W Emulsion 90:10.

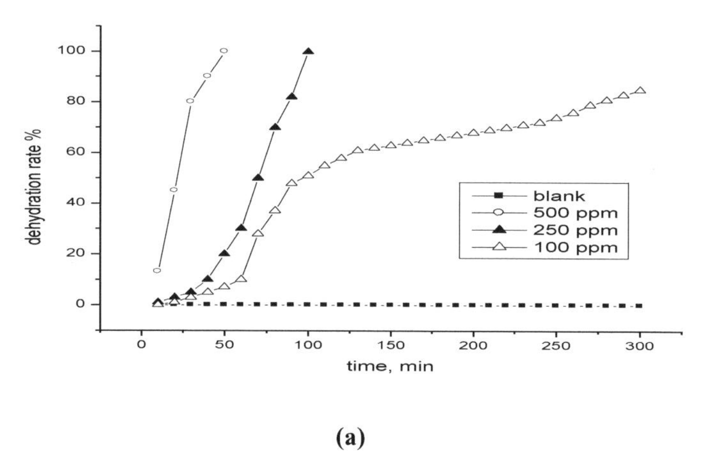

(a)

Fig. (2.21.a): Effect of Concentration on the Dehydration Rate of a) PEG600-PPO1-PEG600 at 333 K for an O:W Emulsion 80:20.

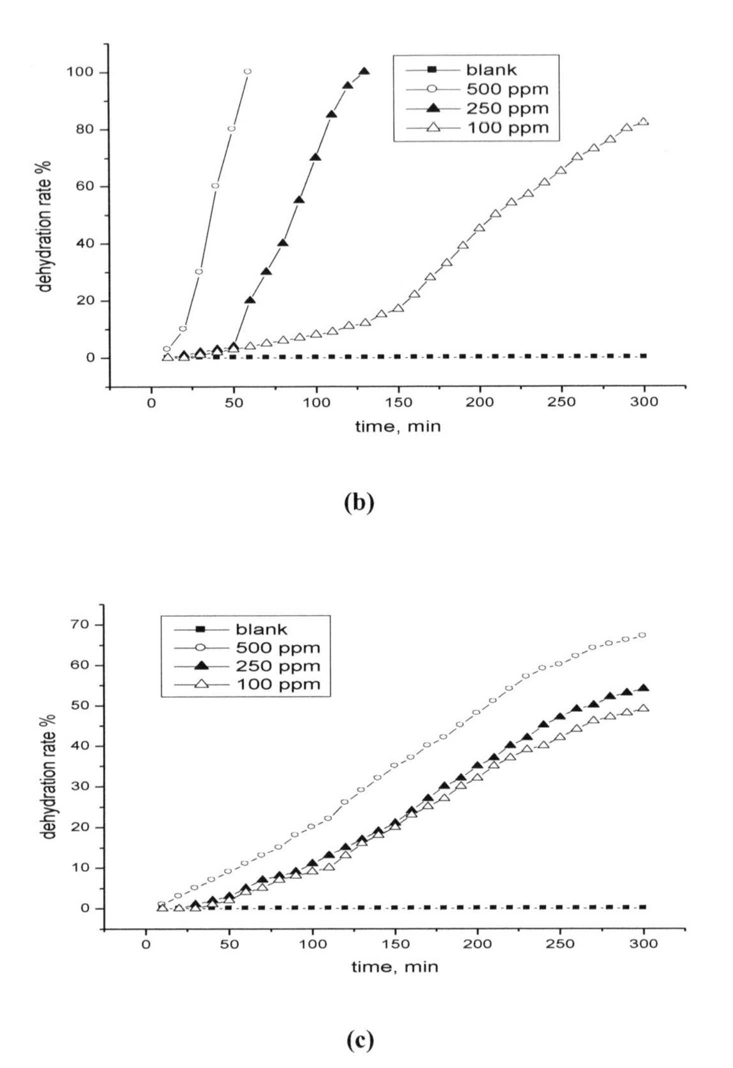

(b)

(c)

Fig. (2.21. b,c): Effect of Concentration on the Dehydration Rate of b) PEG400-PPO1-PEG400 and c) PEG600-PPO2-PEG600 at 333 K for an O:W Emulsion 80:20.

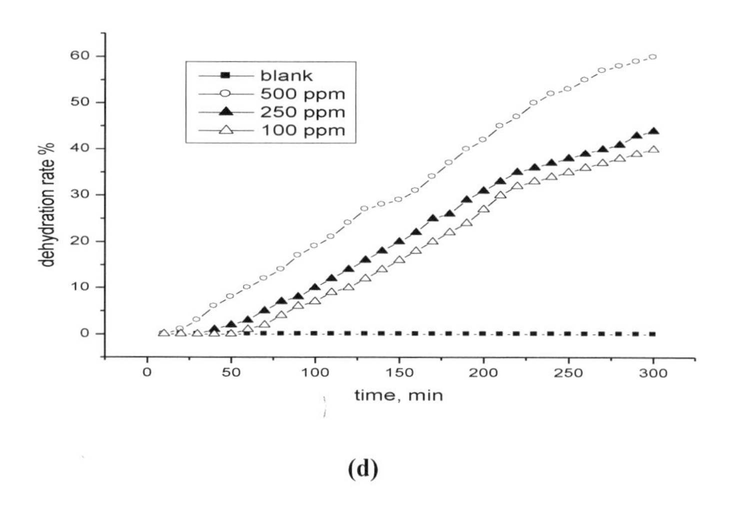

(d)

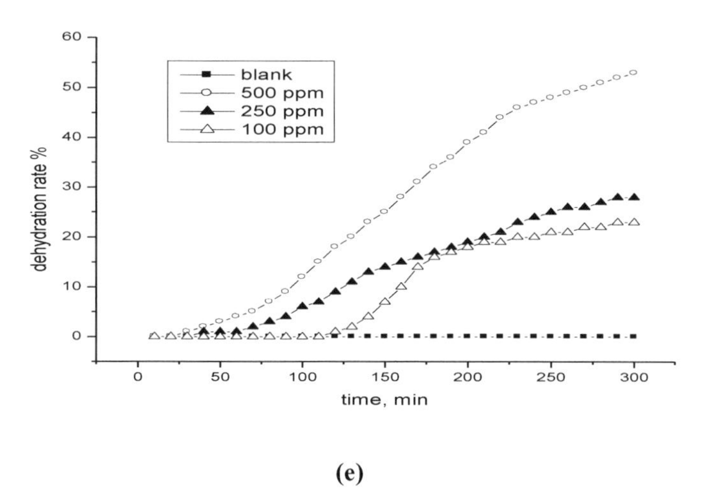

(e)

Fig. (2.21. d,e): Effect of Concentration on the Dehydration Rate of d) PEG1500-PPO2-PEG1500 and e) PEG3000-PPO2-PEG3000 at 333 K for an O:W Emulsion 80:20.

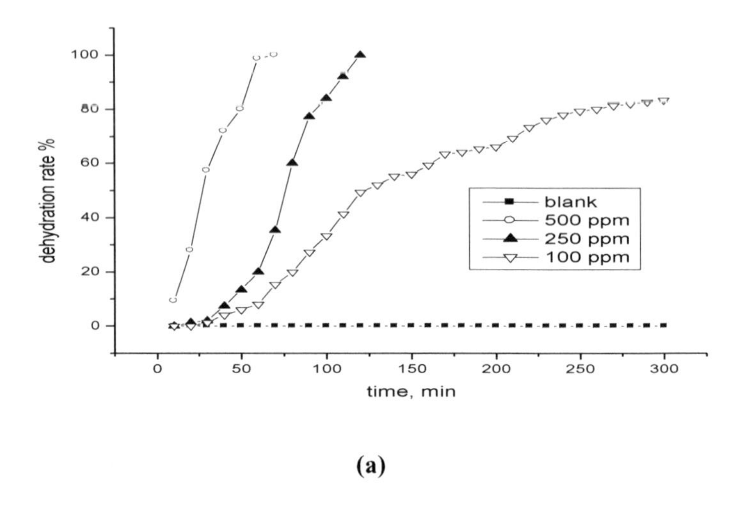

(a)

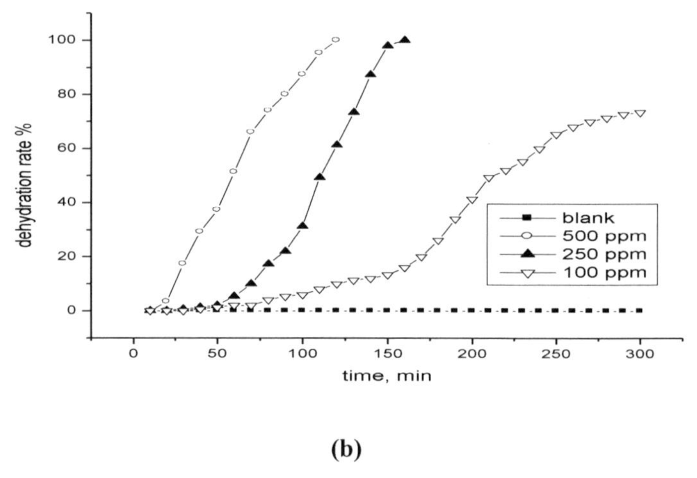

(b)

Fig. (2.22. a,b): Effect of Concentration on the Dehydration Rate of a) PEG600-PPO1-PEG600 and b) PEG400-PPO1-PEG400 at 333 K for an O:W Emulsion 70:30.

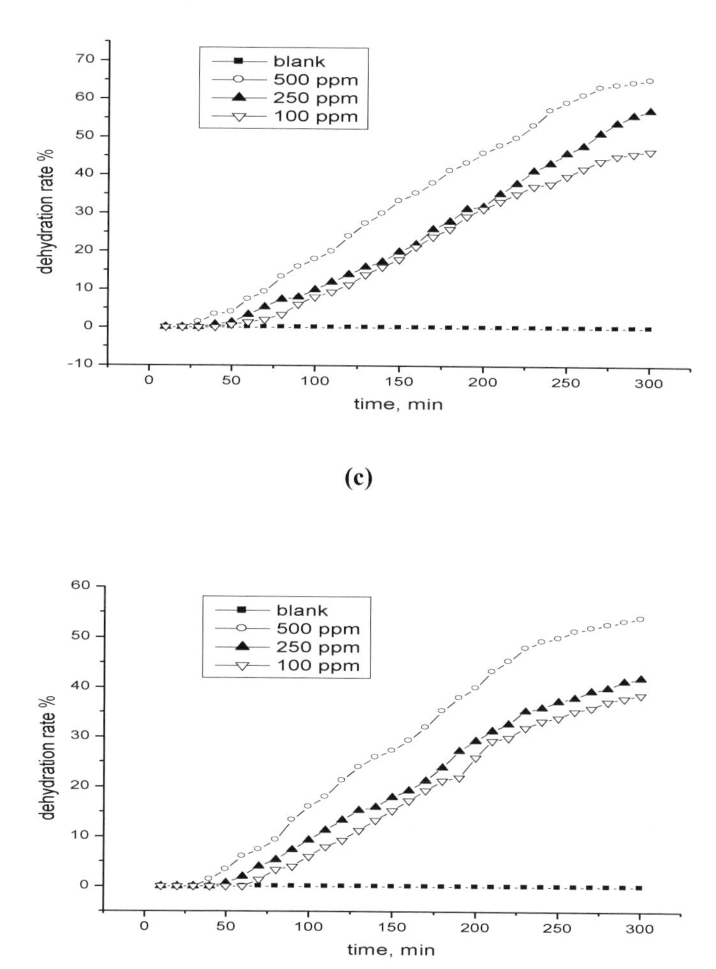

Fig. (2.22. c,d): Effect of Concentration on the Dehydration Rate of c) PEG600-PPO2-PEG600 and d) PEG1500-PPO2-PEG1500 at 333 K for an O:W Emulsion 70:30.

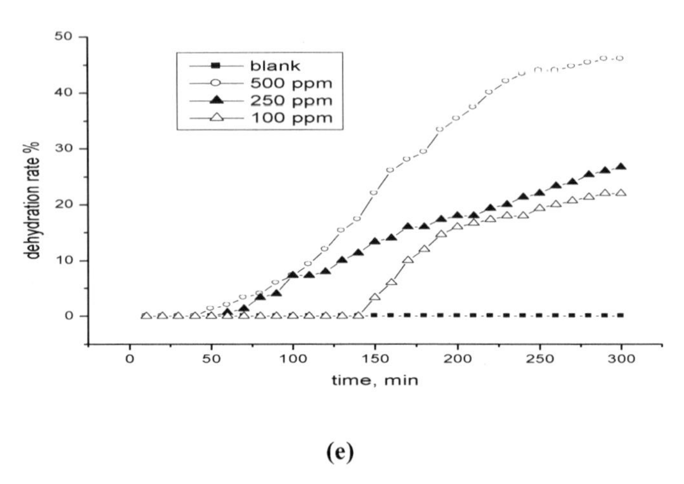

(e)

Fig. (2.22.e): Effect of Concentration on the Dehydration Rate of e) PEG3000-PPO2-PEG3000 at 333 K for an O:W Emulsion 70:30.

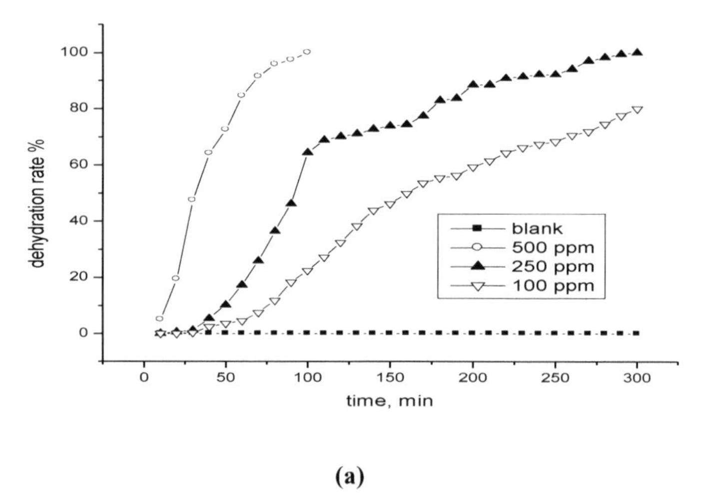

(a)

Fig. (2.23.a): Effect of Concentration on the Dehydration Rate of a) PEG600-PPO1-PEG600 at 333 K for an O:W Emulsion 50:50

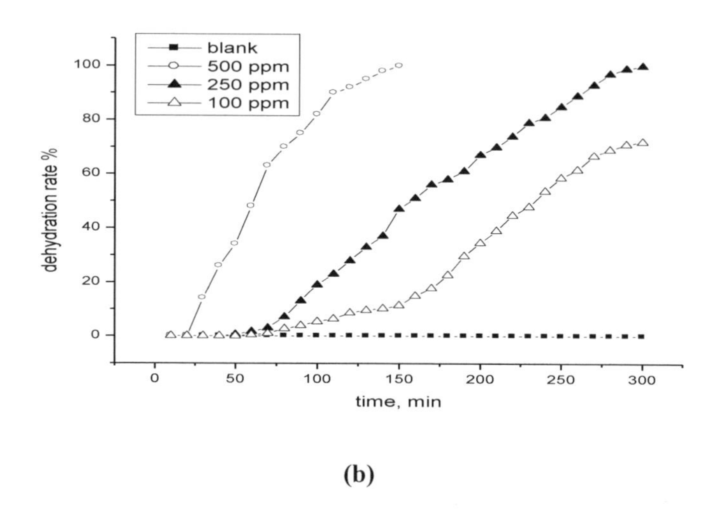

(b)

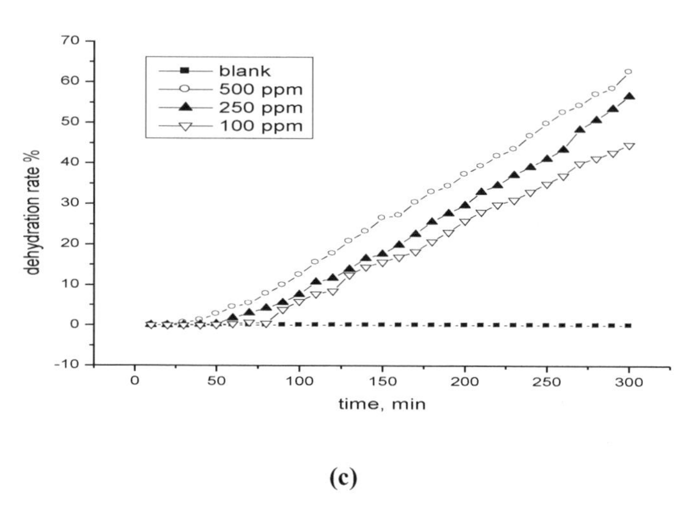

(c)

Fig. (2.23.b,c): Effect of Concentration on the Dehydration Rate of b) PEG400-PPO1-PEG400 and c) PEG600-PPO2-PEG600 at 333 K for an O:W Emulsion 50:50

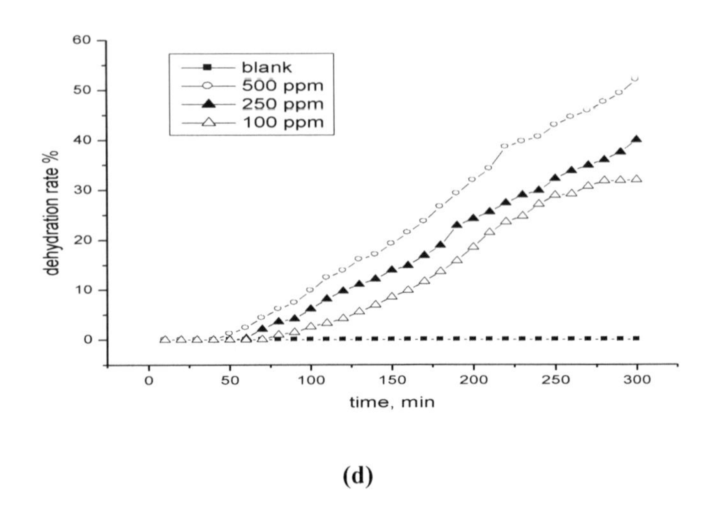

(d)

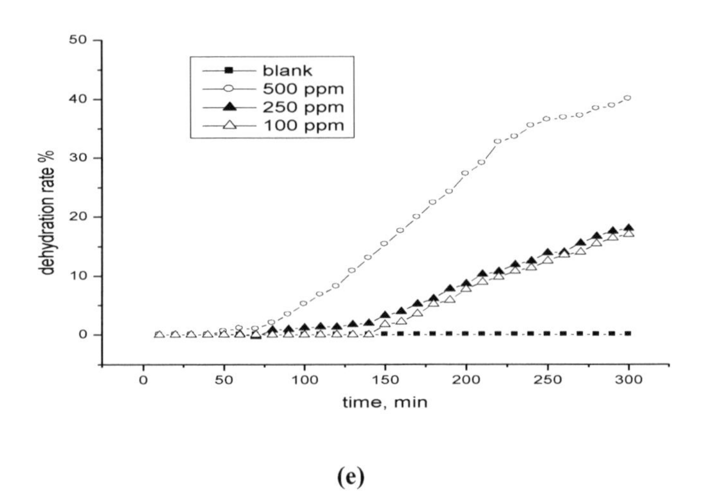

(e)

Fig. (2.23.d,e): Effect of Concentration on the Dehydration Rate of d) PEG1500-PPO2-PEG1500 and e) PEG3000-PPO2-PEG3000 at 333 K for an O:W Emulsion 50:50

(Group 2)

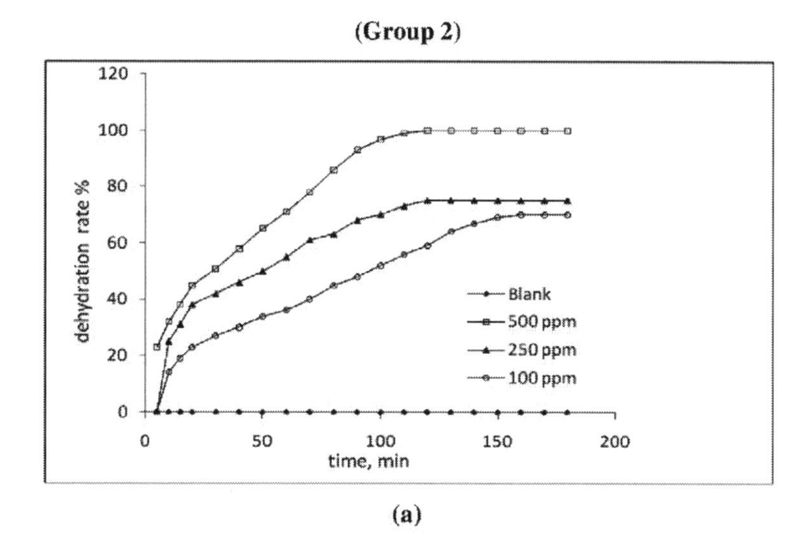

(a)

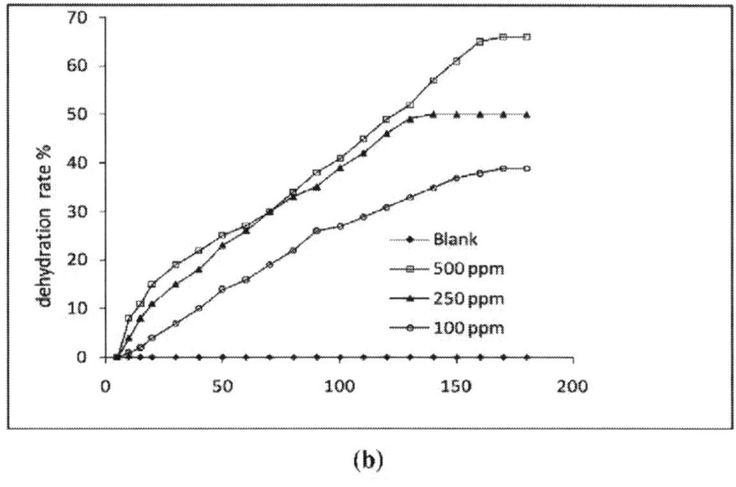

(b)

Fig. (2.24. a,b): Effect of Concentration on the Dehydration Rate of a) PPO-MA(5)-PEG400 and b) PPO-MA(5)-PEG600 at 333 K for an O:W Emulsion 90:10

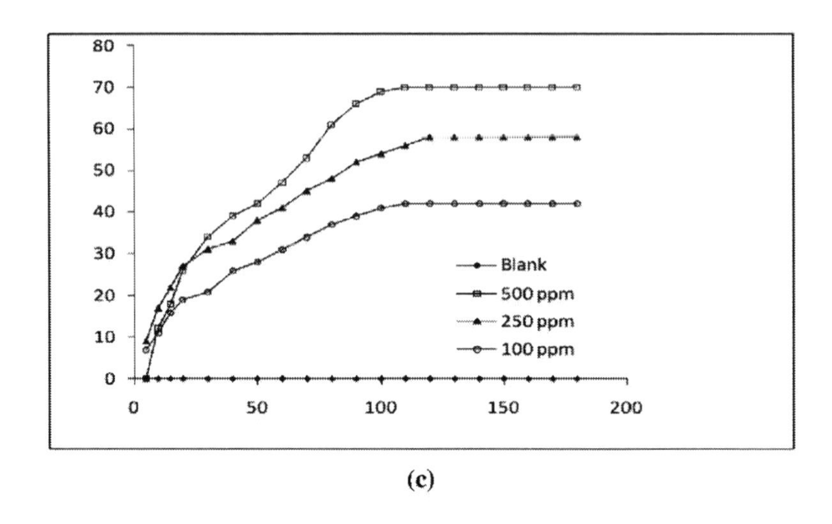

(c)

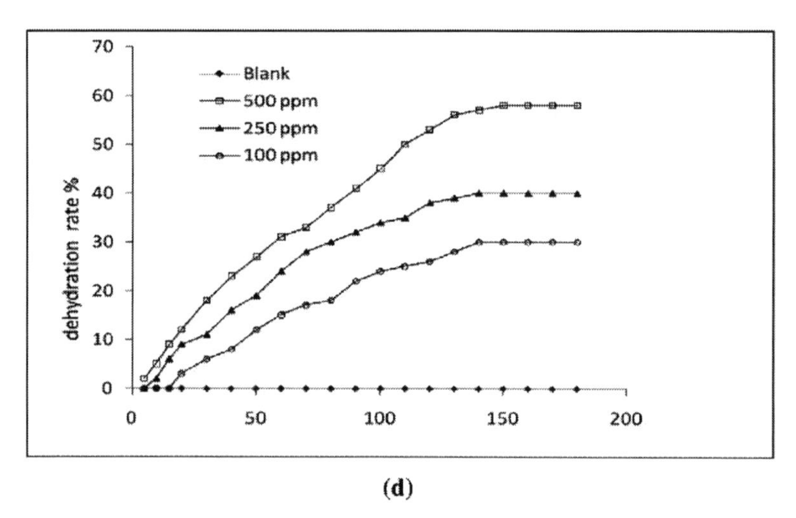

(d)

Fig. (2.24. c,d): Effect of Concentration on the Dehydration Rate of c) PPO-MA(10)-PEG400 and d) PPO-MA(10)-PEG600 at 333 K for an O:W Emulsion 90:10.

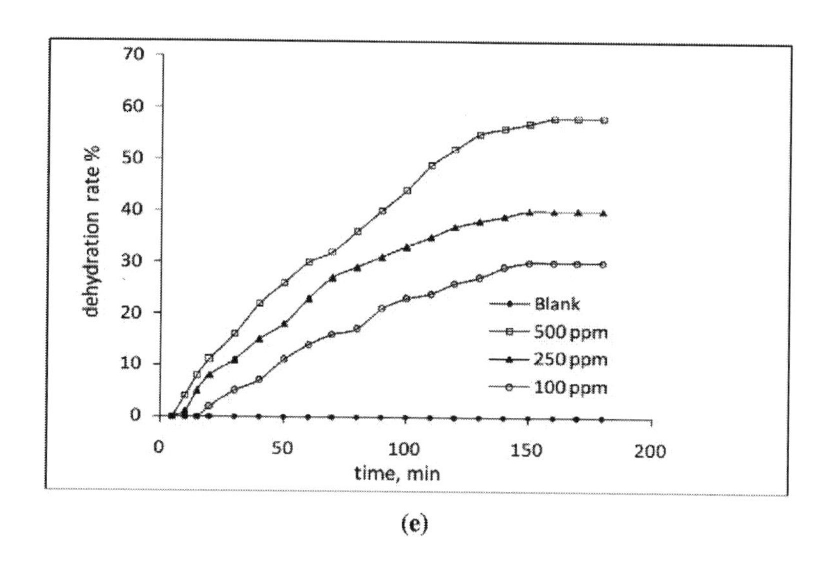

(e)

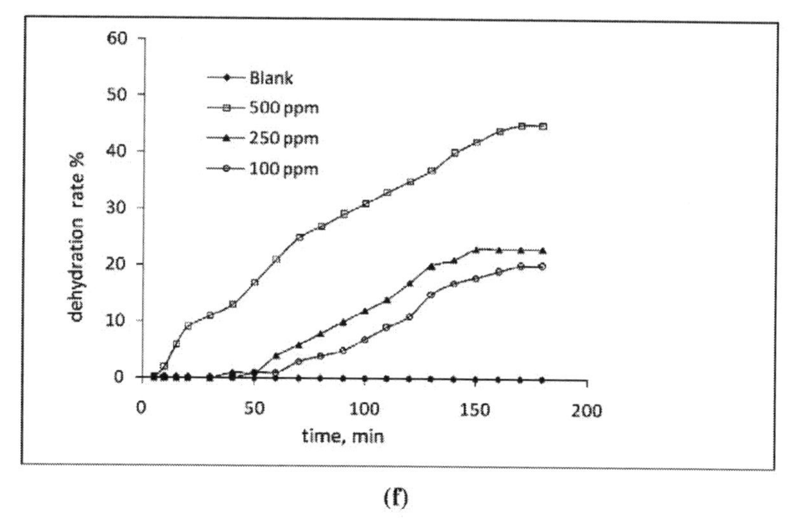

(f)

Fig. (2.24. e,f): Effect of Concentration on the Dehydration Rate of e) PPO-MA(20)-PEG400 and f) PPO-MA(20)-PEG600 at 333 K for an O:W Emulsion 90:10.

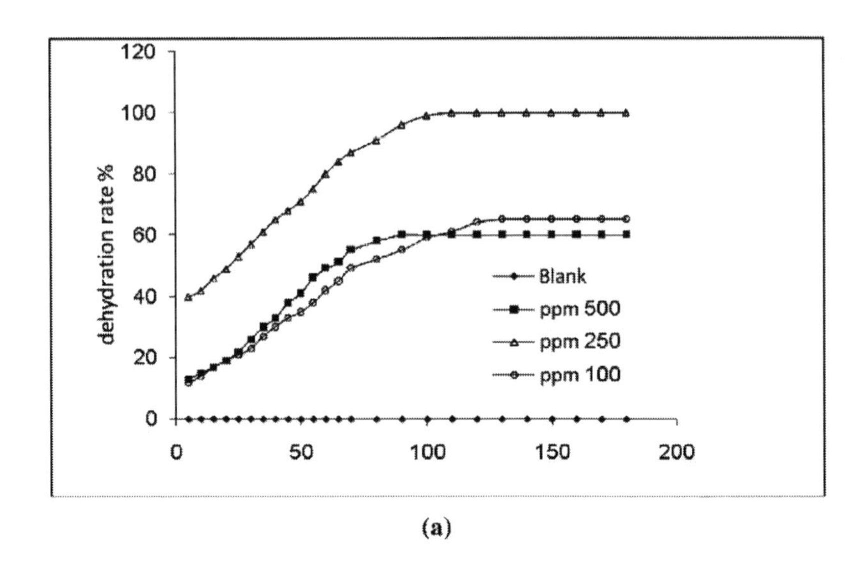

(a)

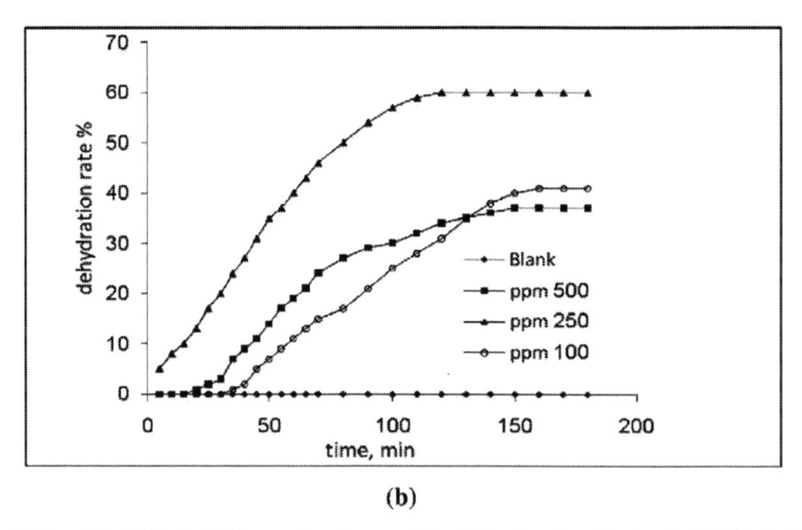

(b)

Fig. (2.25. a,b): Effect of Concentration on the Dehydration Rate of a) PPO-MA(5)-PEG400 and b) PPO-MA(5)-PEG600 at 333 K for an O:W Emulsion 80:20.

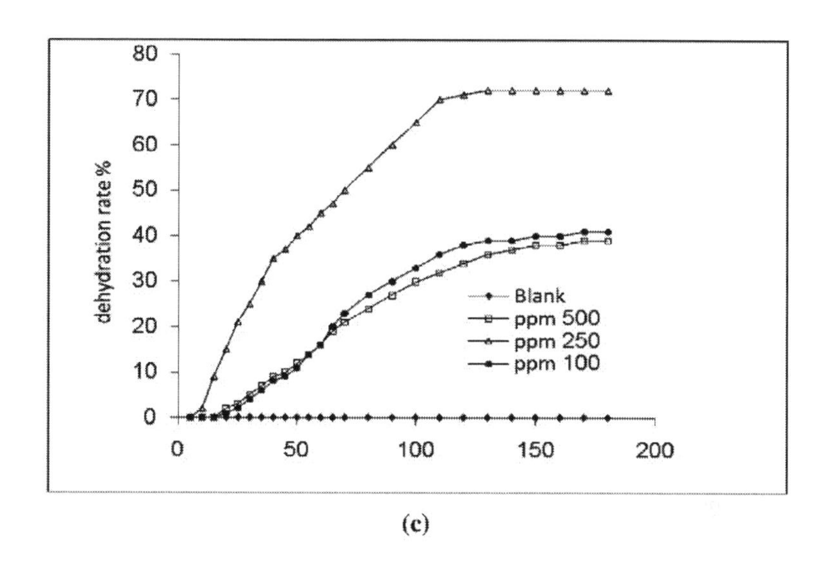

(c)

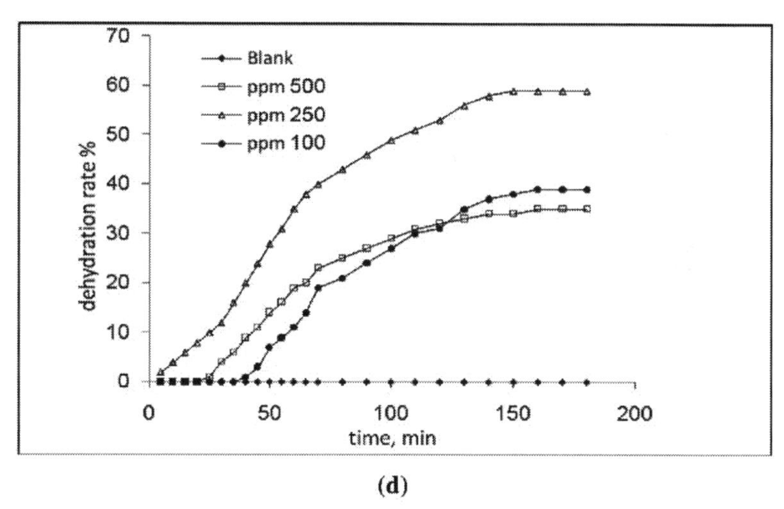

(d)

Fig. (2.25. c,d): Effect of Concentration on the Dehydration Rate of c) PPO-MA(10)-PEG400 and d) PPO-MA(10)-PEG600 at 333 K for an O:W Emulsion 80:20.

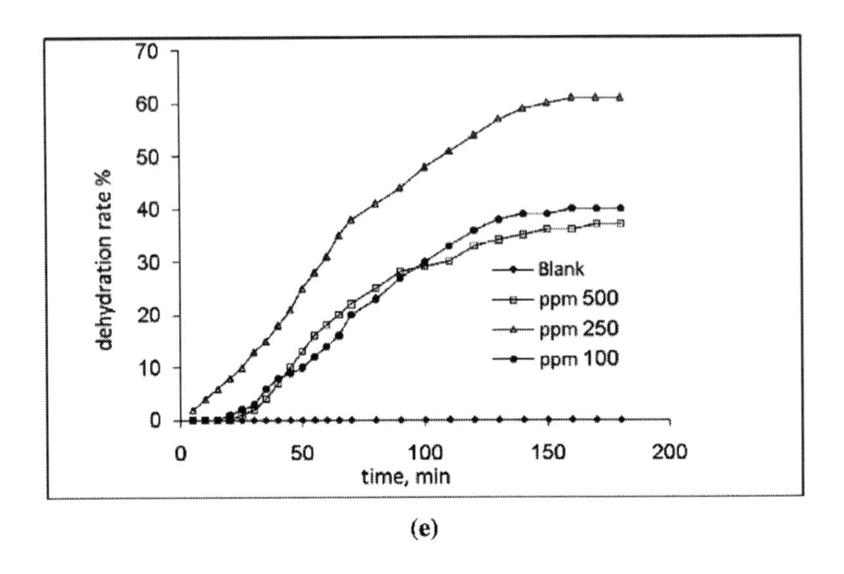

(e)

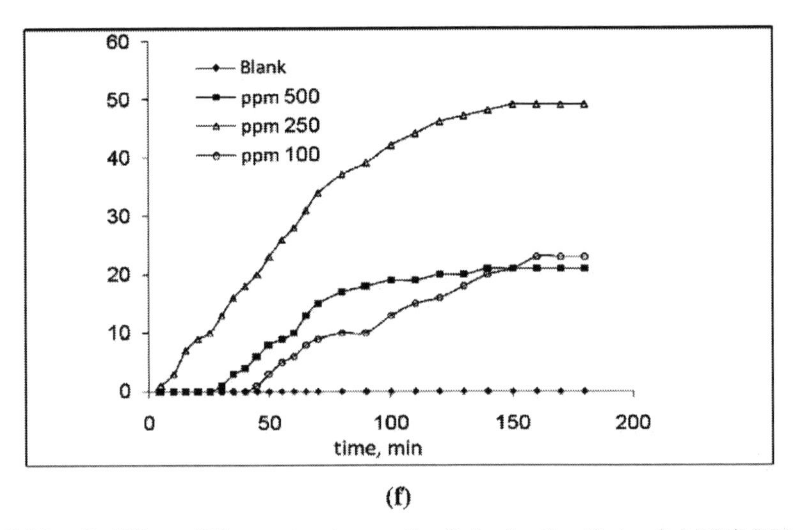

(f)

Fig. (2.25. e,f): Effect of Concentration on the Dehydration Rate of e) PPO-MA(20)-PEG400 and f) PPO-MA(20)-PEG600 at 333 K for an O:W Emulsion 80:20.

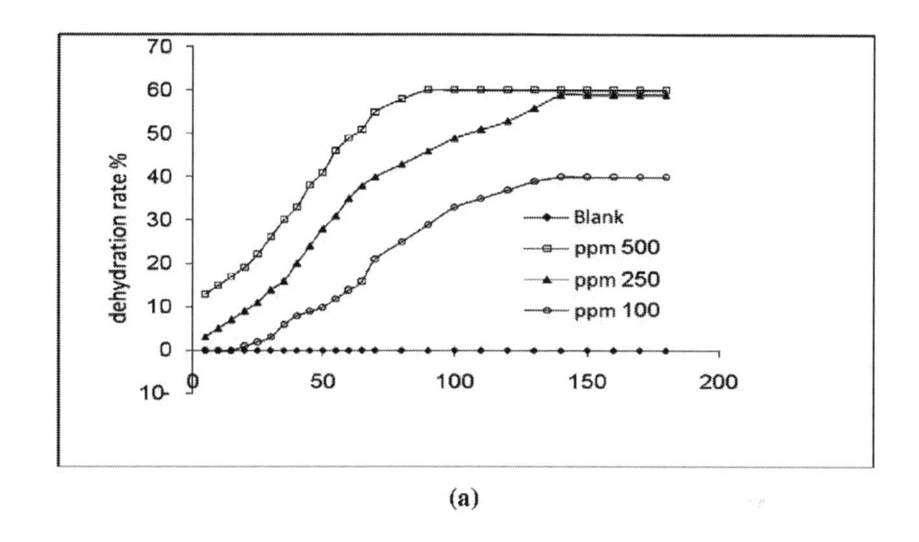

(a)

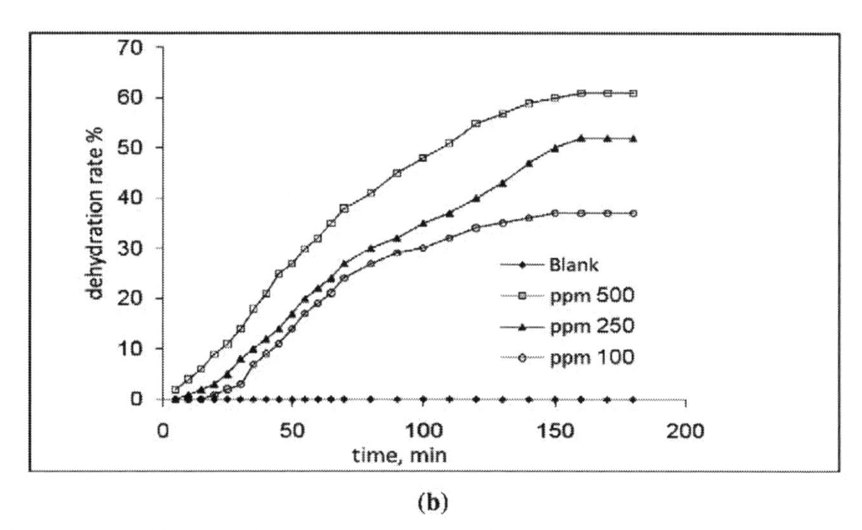

(b)

Fig. (2.26. a,b): Effect of Concentration on the Dehydration Rate of a) PPO-MA(5)-PEG400 and b) PPO-MA(5)-PEG600 at 333 K for an O:W Emulsion 70:30.

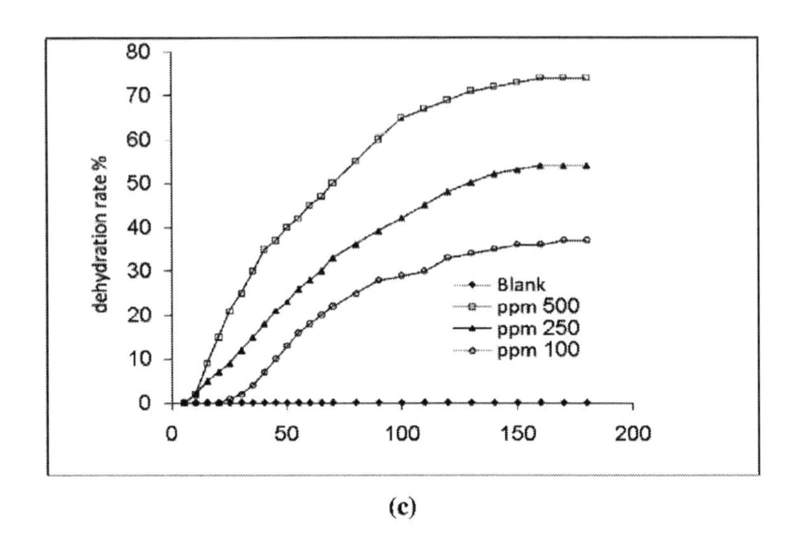

(c)

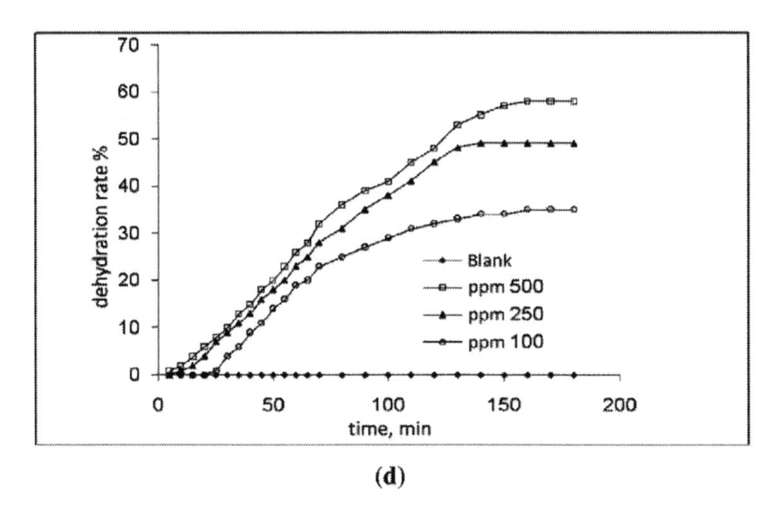

(d)

Fig. (2.26. c,d): Effect of Concentration on the Dehydration Rate of c) PPO-MA(10)-PEG400 and d) PPO-MA(10)-PEG600 at 333 K for an O:W Emulsion 70:30.

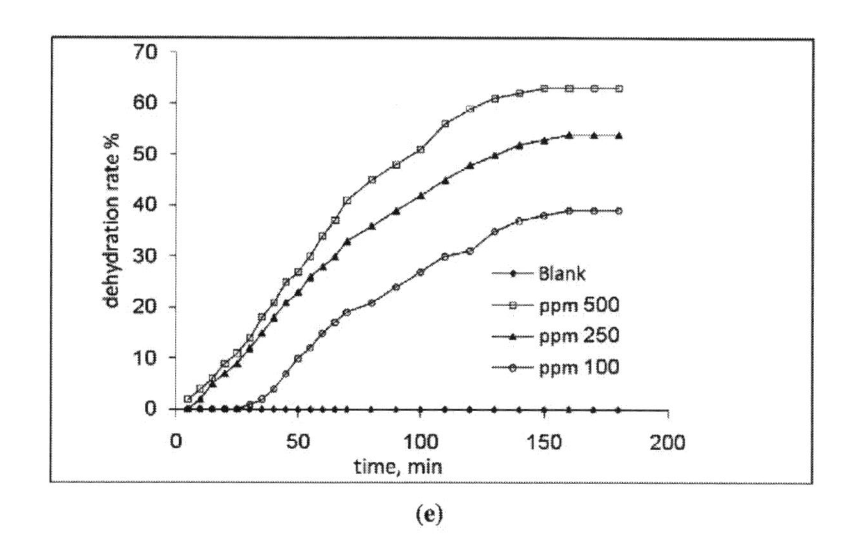

(e)

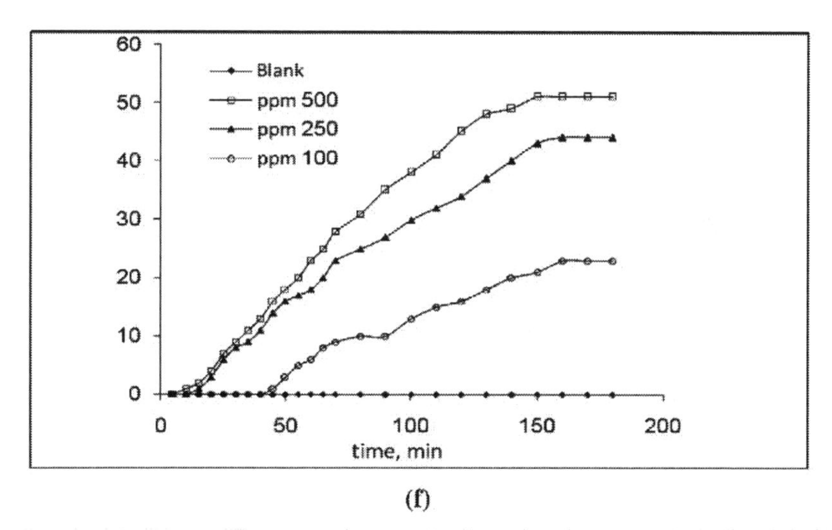

(f)

Fig. (2.26.e,f): Effect of Concentration on the Dehydration Rate of e) PPO-MA(20)-PEG400 and f) PPO-MA(20)-PEG600 at 333 K for an O:W Emulsion 70:30.

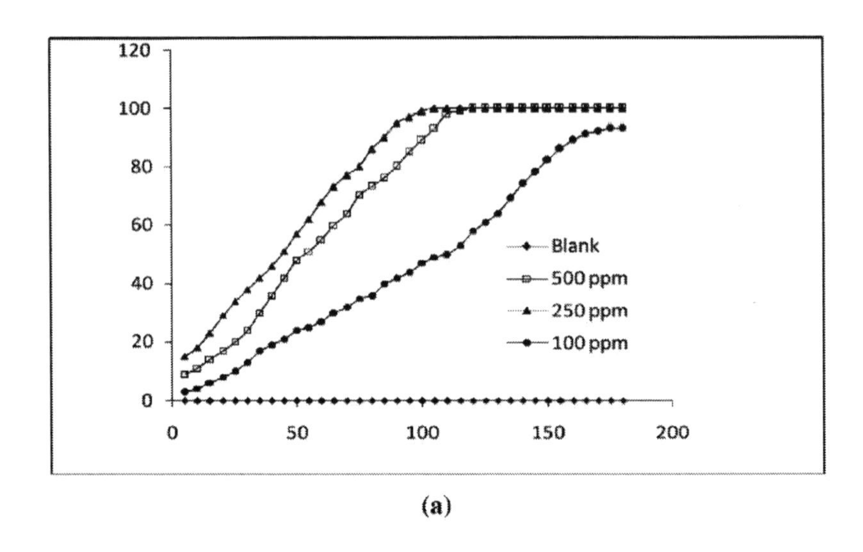

(a)

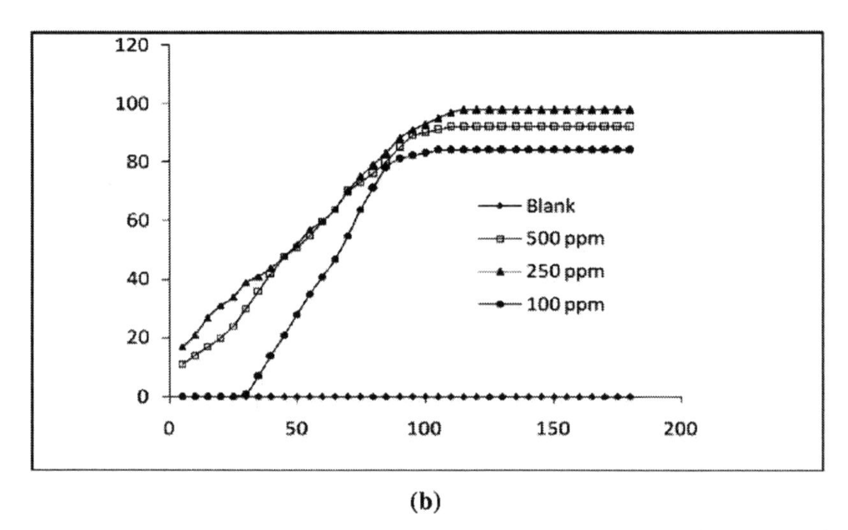

(b)

Fig. (2.27. a,b): Effect of Concentration on the Dehydration Rate of a) PPO-MA(5)-PEG400 and b) PPO-MA(5)-PEG600 at 333 K for an O:W Emulsion 50:50.

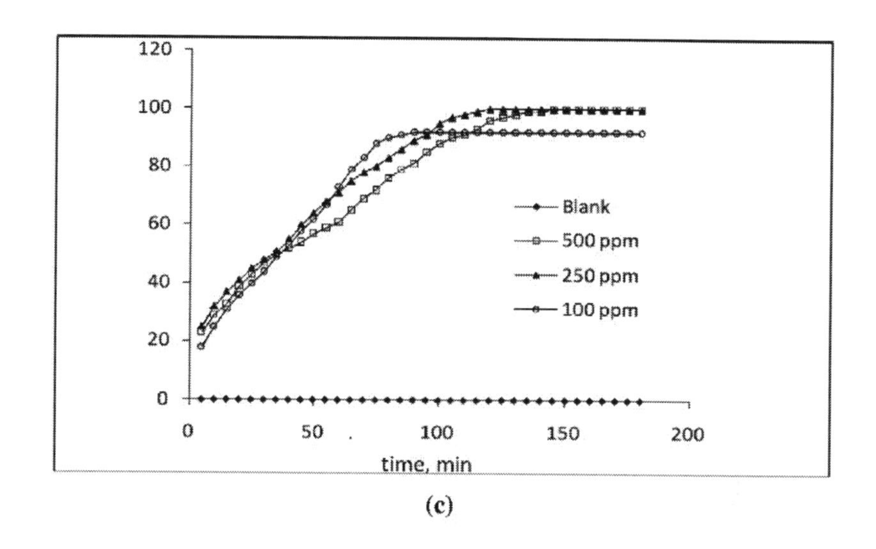

(c)

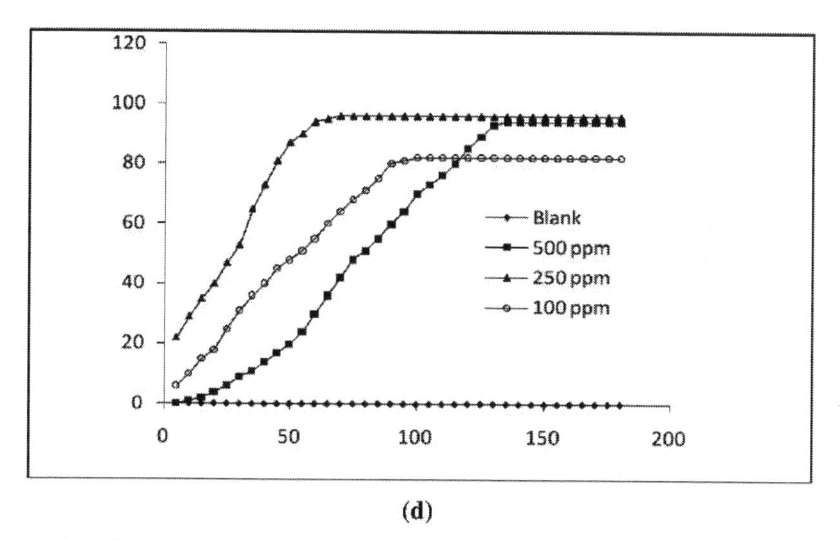

(d)

Fig. (2.27. c,d): Effect of Concentration on the Dehydration Rate of a) PPO-MA(5)-PEG400 and b) PPO-MA(5)-PEG600 at 333 K for an O:W Emulsion 50:50.

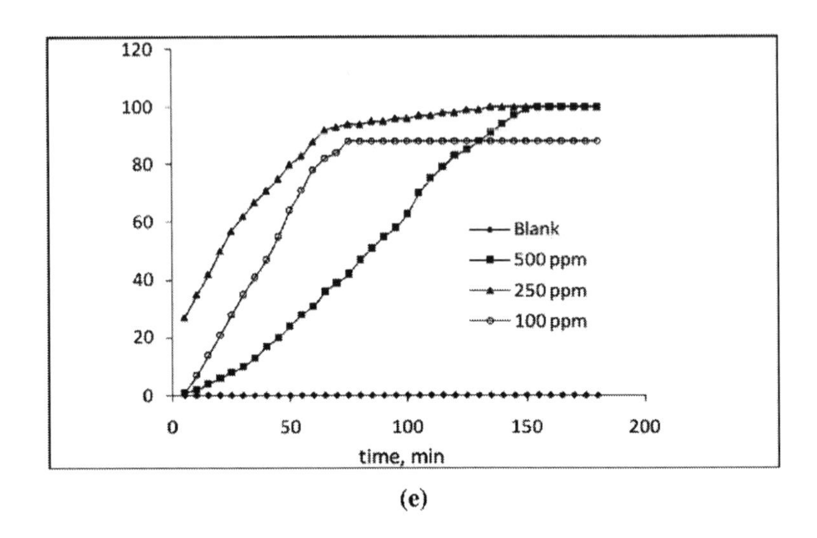

(e)

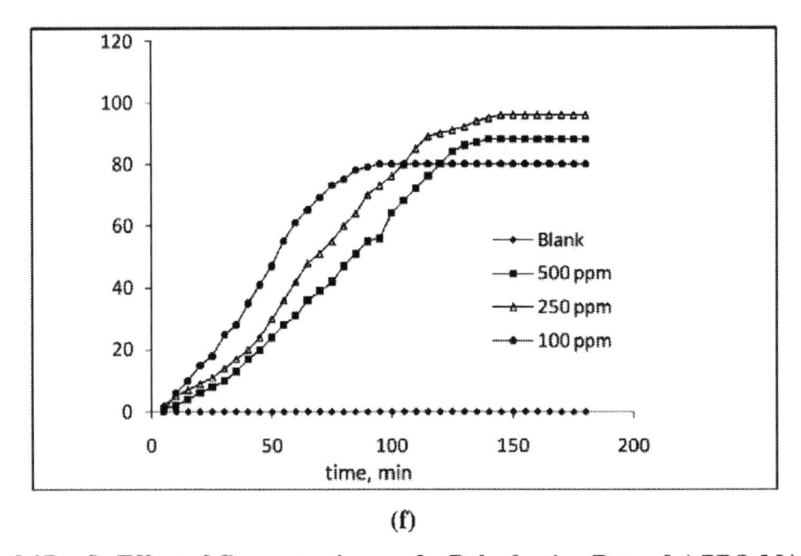

(f)

Fig. (2.27. e,f): Effect of Concentration on the Dehydration Rate of e) PPO-MA(20)-PEG400 and f) PPO-MA(20)-PEG600 at 333 K for an O:W Emulsion 50:50.

CHAPTER IV

"EXPERIMENTAL"

3.1. Materials

a. Propylene glycol (PO)

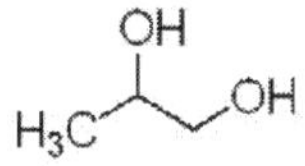

It was obtained from Merck (Germany) with the following specifications:

Molecular formula: $CH_3CH(OH)CH_2OH$

Molecular weight: 76.09

Boling point (b.p.): 187°C

Melting point (m.p.): −60°C

b. Poly (ethylene glycol)400 (PEG 400)

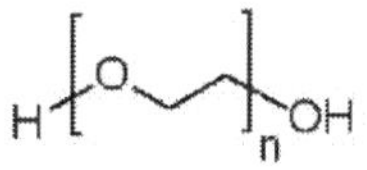

It is obtained from Aldrich Chemical Co. Ltd. (UK) with the following specifications:

Molecular formula: $H(OCH_2CH_2)_nOH$

Molecular weight: 400

Boling point (b.p.): 187 °C

Melting point (m.p.): 4-8 °C

Form: viscous liquid

Viscosity: 210 °F

c. Poly (ethylene glycol)600 (PEG 600)

It is obtained from Aldrich Chemical Co. Ltd. (UK) with the following specifications:

Molecular formula: $H(OCH_2CH_2)_nOH$

Molecular weight: 600

Melting point (m.p.): 20-25 °C

Form: waxy solid (moist)

Density: 1.128 g/ml at 25 °C

Viscosity: 210 °F

d. Poly (ethylene glycol)1500 (PEG 1500)

It is obtained from Aldrich Chemical Co. Ltd. (UK) with the following specifications:

Molecular formula: $H(OCH_2CH_2)_nOH$

Molecular weight: 1500

Melting point (m.p.): 45-50 °C

Autoignition temperature: 581 °F

e. Poly (ethylene glycol)3000 (PEG 3000)

It is obtained from Aldrich Chemical Co. Ltd. (UK) with the following specifications:

Molecular formula: H(OCH2CH2)nOH

Molecular weight: 3000

Melting point (m.p.): 56-59 °C

Autoignition temperature: 581 °F

f. Poly (ethylene glycol) monomethyl ether 400 (PEGME 400)

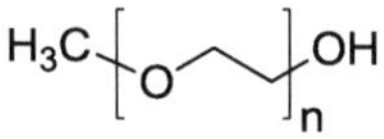

H_3C O OH n

It is obtained from Aldrich Chemical Co. Ltd. (UK) with the following specifications:

Molecular formula: $CH_3(OCH_2CH_2)_nOH$

Molecular weight: 400

Melting point (m.p.): −8 °C

Density: 1.088 g/mL at 20 °C

Description: non-ionic

g. Poly (ethylene glycol) monomethyl ether 600 (PEGME 600)

It is obtained from Aldrich Chemical Co. Ltd. (UK) with the following specifications:

Molecular formula: $CH_3(OCH_2CH_2)_nOH$

Molecular weight: 600

Melting point (m.p.): −8 °C

Density: 1.089 g/mL at 25 °C

Description: non-ionic

Viscosity: 210 °F

Form: semisolid

h. Maleic anhydride (MA)

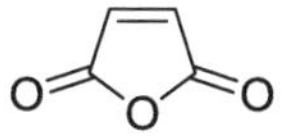

It was obtained from Merck (Germany) with the following specifications:

Molecular formula: $C_4H_2O_3$

Molecular weight: 98.06

Boling point (b.p.): 200 °C

Melting point (m.p.): 51-56 °C

i. Acetic anhydride

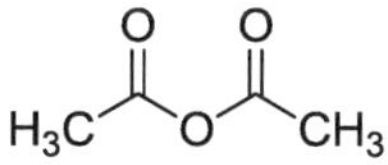

It was obtained from Merck (Germany) with the following specifications:

Molecular formula: $(CH_3CO)_2O$

Molecular weight: 102.09

Boling point (b.p.): 138-140 °C

Melting point (m.p.): −73 °C

Density: 1.08 g/mL

j. Dibenzoyl peroxide (DBP)

It was obtained from Merck (Germany) with the following specifications:

Molecular formula: $(C_6H_5CO)_2O_2$

Molecular weight: 242.23

Melting point (m.p.): 105 °C

k. P-toluene sulfonic acid (PTSA)

It was obtained from Merck (Germany) with the following specifications:

Molecular formula: $CH_3C_6H_4SO_3H\ H_2O$

Molecular weight: 190.22

Melting point (m.p.): 103-106 °C

l. Potassium hydroxide (KOH)

It was obtained from Merck (Germany) with the following specifications:

Molecular formula: KOH

Molecular weight: 56.11

Colour: white

Melting point (m.p.): 361 °C

The Baker crude oil (produced from General Petroleum Co., Egypt) and its general specifications are listed in Table (3.1). On the other hand, the used sea water was obtained from the Mediterranean Sea, Alexandria, Egypt and its physicochemical characterization is shown in Table (3.2).

The solvents used in this study and their sources are listed in Table (3.3).

Distilled water used in surface tension measurements.

Table (3.1): Specifications of baker crude oil

Test	Method	Value
API gravity at 60 F	Calculated	21.7
Viscosity at 60 F (Cst)	IP71	762.8
Specific gravity at 60 F	IP 160/87	0.843
Asphaltene Contents (WT%)	IP 143/84	7.83

Table (3.2): General characterizations of sea-water

Total dissolved solids	44372 mg/l	pH	7.74 at 19 °C
Resistivity	0.019 Ohm at 19 °C	Salinity	39996 mg/l
Conductivity	2.2 mS/M at 19 °C	Sp. Gr.	1.03304
Density	1.032 g/ml		

Table (3.3): Used solvents throughout the investigation

Used Solvents	Source
Ethyl alcohol absolute	El Nasr Pharmaceutical Chemicals Co., Egypt
HCl	El Nasr Pharmaceutical Chemicals Co., Egypt
Xylene	El Nasr Pharmaceutical Chemicals Co., Egypt
N-Hexane	El Nasr Pharmaceutical Chemicals Co., Egypt

3.2. Preparation of PPO

Poly (propylene glycol), PPO was prepared through anion polymerization of PG and propylene oxide (PO) in the presence of 20 (wt %) of KOH as a catalyst under N_2 atmosphere. The temperature of reaction was kept constant at 298 k for 50 hours. The reaction mixture was agitated during the reaction time by adjusting the PO monomer feeding rate. The produced polymers were treated with HCl to form neutral solution.

3.3. Preparations of Demulsifiers "Preparation of PEG-PPO-PEG polymers"

3.3.1. By esterfication reaction of PPO

The esterification reactions of PPO were carried out in a reactor fitted with a Dean–Stark separator, mechanical stirrer, and thermometer and nitrogen inlet. PPO (0.1 mol) and 0.1 mol of maleic anhydride (MA) were mixed in a reactor in the presence of *o*-xylene and 0.1 wt% PTSA (based on total weight of reactants). The esterification reaction was carried out under nitrogen atmosphere. The reaction temperature was raised to 393 K until the calculated amount of water was removed from the reaction mixture. The products was then esterified with 0.1 mol of PEG having different molecular weights in the presence of 0.1 wt% PTSA until the theoretical amount of water was removed at 423 K. o-Xylene was distilled off from the reaction product by rotary evaporator under reduced pressure. The product was separated by salting out use saturated NaCl solution and extracted with isopropanol using separating funnel. The purified products were isolated after evaporation of isopropanol.

3.3.2. By grafting of maleic anhydride onto PPO

The acetylation reaction of PPO was carried out in a reactor fitted with a two-nicked flask equipped with a condenser, mechanical stirrer, and thermometer and nitrogen inlet. PPO (1 mol) and 2 mol of acetic anhydride were refluxed in a reactor. The acetylation reaction was carried out under nitrogen atmosphere. The reaction temperature was 353 K for 10 hours. The product was neutralized with 0.1N KOH then precipitate PPO diacetate (PPO-Ac) with ethyle alcohol. The grafting of MA was carried out by adding different weight ratio onto PPO-Ac in presence of dibenzoyl peroxide (DBP) as shown in Table (3.4).

PPO-Ac (gm)	MA concentration Related to PPO-Ac	DBP (gm)
5	5	0.006
5	10	0.012
5	20	0.025

The mixture was heated to 353 K for 10 hours. The MA was removed after distillation under vacuum for 30 minute at 363 K and 1 mbar then the mixture is extracted with n-hexane several times. The esterification reactions of PPO-MA were carried out in a reactor fitted with a Dean–Stark separator, mechanical stirrer, and thermometer and nitrogen inlet with 1 mol of PEGME having different molecular weights in the presence of *o*-xylene and 0.1 wt% PTSA (based on total weight of reactants) until the theoretical amount of water was removed at 393 K. o-Xylene was distilled off from the reaction product by rotary evaporator under reduced pressure. The product was separated by salting out use saturated NaCl solution and extracted with isopropanol using separating funnel. The purified products were isolated after evaporation of isopropanol.

3.4. Preparation of Water in Crude-Oil Emulsions (W/O)

All emulsions were prepared with a total volume of 100 mL. The ratio between crude oil and the aqueous phase (sea water) was varied from 10-50 %(volume %). The emulsions were prepared by mixing using a Silverstone homogenizer. The speed was ca.1500 rpm for 1h. In this respect, in 500 ml beaker, the crude oil was stirred at 308 K (1500 rpm) while sea water was added gradually to the crude oil until the two phases become completely homogenous. The emulsion was pronounced at different ratios of crude oil: water (90:10, 80:20, 70:30 and 50:50).

3.5. Measurements

3.5.1. Nuclear Magnetic Resonance Spectroscopic Analysis ^{1}H NMR

The ^{1}H-NMR spectra of the prepared Schiff base monomers and polymers were recorded on a 400 MHz Bruker Avance DRX-400 spectrometer (Belgium) and at Cairo university using a DPX300 ^{1}H-NMR Spectrometer (300MHz).

3.5.2. Infrared Spectroscopic Analysis IR Spectra

Infrared spectra were determined with a Perkin-Elmer model 1720 FTIR (KBR).

3.5.3. Surface Tension

Water-soluble polymers were subjected to surface tension measurements. Different concentrations of each sample were prepared and the surface tension at 298, 308, 318 and 328K was measured using a platinum plate tensiometer, model Dognon Abribat Prolabo. A specially designed double jacket glass cell connected with a thermostated oil bath was used for maintaining the adjusted temperature.

Doubly distilled water (γ = 72 dyne/cm) was used for preparing the concentrated stock solutions of the grafts. Several concentrations were prepared by diluting the stock solution with doubly distilled water to the appropriate concentration to be used in the determination of critical micelle concentration (CMC). The diluted solutions were allowed to stand for 24 hr before the surface tension measurements were performed.

3.3.6. Interfacial Tension

The interfacial tension (IFT) of emulsions or between crude oil emulsions and chemical solutions at 45°C was determined by Du Nouy ring method using a Kruss K-12 tensiometer.

3.3.7. Cloud Point

Different solutions of graft copolymers having 2 wt % of the polymer in both double distilled water and saline solutions (1-5 wt % NaCl) were prepared. Each solution was heated with stirring until it becomes turbid. Upon cooling, the turbidity starts to disappear. The temperature at which the solution becomes completely clear was recorded as the cloud point of this particular solution.

3.6. Demulsification of the Prepared Emulsions

The bottle test is used to estimate the capability of the prepared demulsifiers in breaking of water in oil emulsions. Demulsification was studied at 318 K and 333 K using gravitational settling using graduated cone-shaped centrifuge tube. The prepared demulsifiers were diluted to 70 % (wt%) using xylene: ethanol mixture (1:1). The concentrations of the demulsifiers were 50, 100, 250 and 500 ppm and were injected into the emulsion using a micropipette. After the contents in the tube had been shaken for 1 min, the tube was placed in a water bath at 318 or 333 K to allow the emulsion to separate. The phase separation was recorded as a function of time. During the settling, the interface between the emulsion and separated water phase can be easily observed. The reason for working at elevated temperatures is to melt the content of wax in the oil and thereby prevent influence from the wax on the emulsion stability. The elevated temperature is also more closely related to the real working temperature used in the processes onshore.

CHAPTER V

"CONCLUSIONS"

The following conclusions can be withdrawn from the previous results in the following points:

1. New water soluble poly (ethylene glycol)–block–poly (propylene oxide)–block–poly (ethylene glycol) and polypropylene glycol-graft-polyethylene glycol copolymers were prepared in the normal condition.
2. The chemical structure of the prepared PEG-PPO-PEG block and graft copolymers was elucidated by IR and ^{1}H NMR analyses.
3. The conversion of all PPO grafting reaction products with maleic anhydride (MA) was approximately 75% of the initial MA concentration. Furthermore, the formed anhydride graft units may be used to carry out further reactions such as esterification with PEGME to form new surfactant molecules with new interesting properties.
4. The data indicated that the percentage of esterification of PEGME was increased with increasing molecular weights of PEGME from 400 to 600 g/mol and with increasing of MA content from 5 to 20 Wt%. This can be attributed to the higher solubility of PPO-MA20 in toluene than PPO-MA5 that enhances the probability for reaction of PPO-MA with PEGME.
5. The surface activity and parameters such as, the critical micelle concentrations (CMC), surface excess concentration Γ_{max}, the minimum area A_{min} at the aqueous-air interface and values of the surfactant effectiveness of surface tension reduction π_{CMC} of the prepared surfactants were determined at different temperatures.
6. It was found that increasing the length of hydrophobic PPO increases the surface excess of molecule and consequently, decreases A_{min} of molecule at air/water interface.
7. The adsorption of the surfactant molecules at air/water interface increase with decreasing length of PEG in the prepared surfactants.

8. The prepared surfactants reduced both surface and interfacial tension of water and water/oil interface.
9. The water separation rate of the PPO-graft- PEGME surfactants was decreased as compared with PEG-PPO-PEG surfactant.
10. The prepared surfactants succeeded to break water/ oil emulsions and their efficiencies were changed by variation water content and temperature of separation.
11. The time of maximum demulsification efficiency was varied from 30 minute up to 300 minutes. The data indicated that the best HLB values for dehydration of the crude oil emulsion are ranged from 11.1 to 12.1.

The results obtained from this work achieved the main goal of this study to prepare different molecular weight polymeric surfactants for treating different types of water-in-crude oil emulsions.

CHAPTER VI

“REFERENCES”

[1] Mingyuan, L.; Mingjin, X., Ma, Y.; Wu, Z.; Christy, A.A.; Colloids Surf. A (2002), 197, 193-201.

[2] Moran, K.; Yeung, A.; Massliyah, J. J. Chem. Eng. Sci. (2006), 61, 6016-6028.

[3] Ekott, E. J.; Akpabio, E. J. J. Eng. Appl. Sci. (2010), 5, 447-452.

[4] Poteau, S.; Argillier, J.; Energy & Fuels (2005), 19, 1337-1341.

[5] Khristov, K.; Taylor, S.D.; Masliyah, H. J. Colloids Surf. A (2000), 174, 183.

[6] Little, R.; Fuel (1974), 53(4), 246.

[7] Stsrk, J.L.; Asomaning, S.; Energy Fuels (2005), 19(3), 1342.

[8] Kilpatrick, P.K.; Spiecker, M.P.; Encylopedic Handbook of Emulsion Technology; Sjoblom, J. (ed.); Marcel Dekker: New York (2002).

[9] Kim, B.Y.; Moon, J.H.; Sung, T.H.; Yang, S.M.; Kim, J.D.; Separation Sci. & Tech. (2002), 37(6), 1307-1320.

[10] Xia, L.; Lu, S.; Cao, G.; J. Colloid Interface Sci. (2004), 271, 504-506.

[11] Al-Sabagh, A.M.; Maysour, N.E.; Naser, N.M.; Noor El-Din, M.R.; J. Dispersion Sci. & Tech. (2007), 28, 537–545.

[12] Morrel, J.C.; Egloff, G.; Colloid Chemistry; Alexander, J. (ed.); Reinhold Publishing Corp.: New York (1946).

[13] Denga, S.; Yu, G.; Jiang, Z.; Zhang, R.; Ting, Y. P.; Colloids Surf. A (2005), 252, 113–119.

[14] Kim, Y.H.; Wasan, D.T.; Ind. Eng. Chem. Res. (1996), 35(4), 1141.

[15] Wang, Y.; Zhang, L.; Sun, T.; Zhao, S.; Yu, J.; J. Colloid & Interface Sci. (2004), 270, 163–170.

[16] Shetty, C.S.; Nikolov, A.D.; Wasan, D.T. J. Dispersion Sci. & Tech. (1992), 13(2), 121-133.

[17] Zhang, Z.; Xu, G.; Wang, F.; Dong, S.; Chen, Y.; J. Colloid Interface Sci. (2005), 282, 1–4.

[18] Mankovich, M. J. Phys. Chem. (1954), 58, 1028.

[19] Heydegger, H.; Dunning, H. J. Phys. Chem. (1959), 63, 1613.

[20] Hergeth, W.; Zimmermann, R.; Bloss, P.; Schmutzler, K.; Wartewing, S. Colloids Surf. (1991) 56, 177.

[21] Diakova, B.; Kaisheva, M.; Platikanov, D.; Colloids & Surfaces A: Physicochem. Eng. Aspects (2001), 190, 61–70.

[22] Farn, R. J. "Chemistry and Technology of Surfactants" Blackwell Publ. Oxford (2006).

[23] Piirma, I. "Polymeric Surfactants" Surfactant Science Series, (1992), 42, Marcel Dekker, NY.

[24] Myers, D. "Surfactant Science and Technology" New York, N.Y.: VCH (1988).

[25] Nagarajan, R. Applied Sciences, Dordrecht: Kluwer Academic Publisher (1996), 327, 121–65.

[26] Butt, H.; Graf, K.; Kappl, M.; Physics and Chemistry of Interfaces, Wiley-VCHVerlag GmbH and Co., Weinheim, 2003, Chapter 1.

[27] Bartell, F.E.; Zuidema, H.H.; J. Am. Chem. Soc. (1936), 58, 1449–1454.

[28] Ali, K.; Anwar-ul-Haq, Salma, B.; Siddiqi, S. Colloids Surf. A (2006), 272, 105–110.

[29] Lee, B.-B.; Ravindra, P.; Chan, E.-S.; Colloids and Surfaces A: Physicochem. Eng. Aspects (2009), 332, 112–120.

[30] Schramm, L.L. "Surfactants; Fundamentals and Applications in the Petroleum Industry" Cambridge Univ. Press, U.K. (2000).

[31] Rosen, M. J. "Surfactants and Interfacial Phenomena", 2nd Ed, Wiley: New York, (1989).

[32] Tadros, Th.F.; Vandamme, A.; Levecke, B.; Booten, K.; Stevens, C.V. Adv. Colloid Interface Sci. (2004), 108-109, 207-226.

[33] Hoeve, C.A. J. Polym. Sci. (1970) 30, 361.

[34] Surfactant operation and effects "Surfactant-Wikipedia, the free Encyclopedia", available online.

[35] Schramm, L.L. “Emulsions; Fundamentals and Applications in the Petroleum Industry” American Chemical Society: Washington, DC (1992).

[36] Kelvens, H. B. J. Am. Oil Chem. Soc. (1953), 30, 74.

[37] Shinoda, K. “Colloidal Surfactants” Academic Press: New York (1963).

[38] Shinoda K.; Friberg, S. “Emulsion and Solubilization” John Wiley, New York (1986).

[39] Griffin, W.C. J. Soc. Cosmet. Chem. (1949), 1, 311.

[40] Flockhart, B. D. J. Colloid Interface Sci. (1961), 16, 484.

[41] Cook, E. H.; Fordyce, D. B.; Trebbi, G. F. J. Phys. Chem. (1963), 67, 1987.

[42] Roberts, M. J.; Bentley, M. D. "Chemistry for peptide and protein PEGylation" Adv. Drug Delivery Rev. (2002), 54, 459-476.

[43] O’Connor, S. M., Gehrke, S. H. Langmuir (1999), 15, 2580-2585.

[44] Diakova, B.; Plantikanov, D.; Atanassov, R.; Kaisheva, M.; Adv. Colloid & Interface Sci. (2003), 104, 25–36.

[45] Wu, J.; Xu, Y.; Dabros T.; Hamza, H. Energy & Fuels, (2003), 17(6) 1554-1559.

[46] Okazaki, M.; Washio, I.; Shibasaki, Y.; Ueda, M. J. Am. Chem. Soc. (2003), 125 (27), 8120–8121.

[47] Hou, S.; El Chaikof, D. Taton; Gnanou, Y. Macromolecules (2003), 36, 3874.

[48] Riess, G. Prog. Polym. Sci. (2003), 28, 1107–1170.

[49] Okano, T.; Ikemi, M.; Shinohara, I. Nippon Kagaku Kaishi (1977) 1702; C.A. (1978), 88, 38240h.

[50] Ruckenstein, E.; Zhang, H. J. Polym. Sci. Part A: Polym.Chem. (2000), 38, 1195.

[51] Tousaint, W.J.; Fife, H.R. U.S. Patent (1977), No. 2, 425, 845.

[52] Akhmedov, U.K.; Niyazova, M.M.; Akhmedov, K.S. Uzb. Khim. Zh, 6, 29-31 (1982); C.A. (1983), 98, 132836k.

[53] Molliet, J.L.; Collie, B.; Black, W. "Surface Activity" 2nd Ed, Spon Ltd., London (1961).

[54] Omar, A. M.; Abdel-Khalek Tenside Surf. Det. (1997), 34 (3), 179.

[55] Bo, G.; Wesslen, B.; Wesslen, K.; J. Polym. Sci., Part A; Polym. Chem. (1992) 30, 1799.

[56] Jackson, D.R.; Lundsted., L.G.; U.S. Patent, No.2, 677,700 (1954); C.A. (1954), 48, 9727g.

[57] Lundsted, L.G.; Schmolka, I.R. "The synthesis and properties of Block copolymer polyol surfactants" and "The applications of Block Copolymer Polyol Surfactants in block and graft copolymerization" Vol. 2, Ceresa, R.J. Ed., John Wiley and Sons, London (1976).

[58] Nakamura, M.; Sasaki, T. Bull. Chem. Soc. (1970), 43, 3667.

[59] Wesslen, B.; Wesslen, K. B. J. Polym. Sci. Part A:Polym. Chem. (1989), 27, 3915.

[60] Velichkova, R. S. and Christova, D. C., Prog. Polymer Sci. (1995), 20, 819-887.

[61] Lee, J. H.; Ju, Y. M.; Kim, D. M. Biomaterials (2000), 21, 683-691.

[62] Zhang, Z.; Xu, G.; Wang, F.; Dong, S.; Chen, Y.; Key Laboratory of Colloid and Interface Chemistry (Shandong University), Ministry of Education, Jinan 250100, People's Republic of China (2004).

[63] Svensson, M.; Johansson, H.; Tjerneld, F. Amphiphilic Block Copolymers, (2000), 377-407.

[64] Wu, J.; Xu, Y.; Dabros, T.; Hamza, H. Canmet Energy Technology Centre – Devon, Natural Resources Canada, 1 Oil Patch Drive, Devon, Alta., Canada T9G 1A82004.

[65] Chiappetta, D. A.; Sosnik, A.; Eur. J. Pharm. & Biopharm. (2007), 66, 303-317.

[66] Arun, A.; Gaymans, R.J. Eur. Polymer J. (2009), 45, 2858-2866.

[67] Malik, M.I.; Trathnigg, B.; Saf, R. J. Chromatography A (2009) 1216, 6627-6635.

[68] Maciejczek, A.; Mass, V.; Rode, K.; Pasch, H. Polymer (2010), 51(26), 6140-6150.

[69] Alexandridis, P.; Hatton, T. A. Colloids and Surfaces A: Physicochemic and Engineering Aspects (1995), 96, 1-46.

[70] Alexandridis, P. Current Opinion in Colloid & Interface Science (1997), 2, 478-489.

[71] Zhang, Z.; Xu, G. Y.; Wang, F.; Dong, S. L.; Li, Y. M. J. Coll. & Interface Sci. (2004), 277(2), 464-470.

[72] Sosnik, A.; Cohn, D. Biomaterials (2005), 26(4), 349-357.

[73] Bakshi, M. S.; Sachar, S.; Yoshimura, T.; Esumi, K. J. Coll. & Interface Sci. (2004), 278(1), 224-233.

[74] Lucie M.A. van de Steeg, Carl-Gustaf Gölander, Coll. & Surf. (1991), 55, 105-119.

[75] Liaw, J.; Lin, Y. J. Controlled Release (2000), 68(2) 273-282.

[76] Gosa, K.; Uricanu, V.; Colloids and Surfaces A: Physicochem. & Eng. Aspects (2002), 197(1-3), 257-269.

[77] Yu-Bing Liou, Ruey-Yug Tsay, Adsorption of PEO–PPO–PEO triblock copolymers on a gold surface, J. Taiwan Instit. Chem. Eng. In Press, Corrected Proof, Available online 4 December 2010.

[78] Gorrasi, G.; Stanzione, M.; Izzo, L.; Reactive & Functional Polymers (2011), 71(1), 23-29.

[89] Husken, D.; Feijen, J.; Gaymans, R. J.; Eur. Polymer J. (2008), 44, 130-143.

[80] Prasad, R. “Petroleum Refining Technology” Khanna Publisher, New Delhy (2000).

[81] Speight, J. G. “The Chemistry and Technology of Petroleum” Marcel Dekker, New York (1999).

[82] Ramos, A. C.; Haraguchi, L.; Notrispe, F. R.; Loh, W.; Mohamed, R. S. J. Petrol. Sci. & Eng. (2001), 32, 201-216.

[83] Andersen, S. I.; Speight, J. G. J. Petrol. Sci. & Eng. (1999), 22(1-3), 53-66.

[84] Al-Sabagh, A. M.; Atta, A. M. J. Chem. Tech. & Biotech. (1999), 74, 1075–1081.

[85] Rocha Junior, L.C.; Ferreira, M.S.; Ramos, A. Carlos da Silva; J. Petrol. Sci. & Eng. (2006), 51, 26-36.

[86] Qin, X.; Wang, P.; Sepehrnoori, K.; Pope, G. A. Ind. Eng. Chem. Res. (2000), 39, 2644-2654.

[87] H Groenzin, Mullins, O. C. J. Phys. Chem. A. (1999), 103, 11237-11245.

[88] Abraham, T.; Christendat, D.; Karan, K.; Xu, Z.; Masliyah, J.; Indust. & Eng. Chem. Res. (2002), 41, 2170-2177.

[89] Speight, J. G. The Chemistry and Technology of Petroleum, Marcel Dekker, New York (1991).

[90] Swanson, J. M.; J. Phys. Chem. (1942), 46, 141-150.

[91] Agrawala, M. H.; Yarranton, W. Ind. Eng. Chem. Res. (2001), 40, 4664-4672.

[92] Sheu, E. Y.; Liang, K. S.; Sinha, S. K.; Over-field, R. E.; J. Colloid Interf. Sci. (1992), 153, 399.

[93] Neumann, H. J.; Paczynska-Lahme, B.; Severin, D.; Composition and Properties of Petroleum, Ferdinand Enke Publishers, Germany (1981).

[94] Peramanu, S.; Pruden, B. B. Ind. Eng. Chem. Res. (1999), 38, 3121-3130.

[95] Mushrush, G. W.; Speight, J. M.; Petroleum Products: Instability and Incompatibility, Taylor & Francis, London (1995).

[96] Spiecker, P.M.; Gawrys, K.L.; Kilpatrick, P. K.; J. Coll. & Interface Sci. (2003), 267, 178-193.

[97] Rogacheva, O. V.; Rimaev, R. N.; Z.Gubaidullin, V.; Khakimov, D. K.; Investigation of the surface activity of the asphaltenes of petroleum residues. Ufa Petroleum Institute [Translated from Kolloidnyi Zhur.] (1980), 42(3), 586.

[98] Priyanto, S.; Mansoori, G.A.; Suwono, A. Chem. Eng. Sci. (2001), 56, 6933–6939.

[99] Espinat, D.; Ravey, J.C. Colloidal structure of asphaltene solutions and heavy-oil fractions studied by small-angle neutron and X-ray scattering. In: SPE International Symposium, New Orleans, LA (1993), Pp. 365-373.

[100] Moschopedis, S. E.; Speight, J. G. Fuel (1976), 55, 187.

[101] Firoozabadi, A. Thermodynamics of Hydrocarbon Reservoirs, McGraw-Hill, New York (1999).

[102] Rogel, E.; Leon, O.; Espidel, J.; Gonzales, Y. "Asphaltene Stability in Crude Oils," SPE Production and Facilities (2001), 16, 84.

[103] Chia-Lu Chang, Scott Fogler, H. Langmuir (1994), 10, 1758–1766.

[104] Koots, J.A.; Speight, J.G. Fuel (1975), 54, 179.

[105] Groenzin, H.; Mullins, O. C. Schlumberger-Doll Research, Ridgefield, Connecticut 06877, Energy Fuels, (2000), 14, 677–684.

[106] Philp, R.P.; Hseih, M.; Tahira, F.; A review of developments related to the characterization and significance of high molecular weight paraffins (>C40) in crude oils. In: Cubitt, J.M.; England, W.A.; Larter, S. (Eds.). Understanding Petroleum Reservoirs: Towards an Integrated Reservoir Engineering and Geological Approach. Geological Society Special Publication, (2004), 237: 37 – 51.

[107] Jorda, R.M. Paraffin Deposition and Prevention in Oil Wells, JPT 1605, Trans, AIME 237 (1966).

[108] Tuttle, R.N. J. Pet. Tech. (1983), 1192-1196.

[109] Sams, G. W.; Zaouk, M.; Energy & Fuels (2000), 14, 31-37.

[110] Lee, J.M.; Lim, K.H.; Smith, D.H. Langmuir (2002), 18, 7334–7340.

[111] Yan, N.; Kurbis, C.; Masliyah, J.H. Am. Chem. Soc. Ind. Eng. Chem. Res. (1997), 36, 2634–2640.

[112] Schubert, H.; Armbroster, H.; Intl. Chem. Eng. (1992), 32, 14-28.

[113] Kokal, S.L. "Crude-Oil Emulsion" Petroleum Engineering Hand-book, SPE, Richardson, Texas (2005).

[114] McLean, J.D.; Kilpatrick, P.K. J. Colloid Interface Sci. (1997), 189, 242–253.

[115] Zaki, N.; Schorling, P. C.; Rahimian, I. Pet. Sci. Technol. (2000), 18, 945–963.

[116] Harris, J.R.; Hydrocarbon Process (1996), 75(8), 63-68.

[117] Khatib, Z.; Verbeek, P. J. Petrol. Technsol. (2003), 55, 26.

[118] Denga, S.; Yub, G.; Jiangb, Z.; Zhangc, R.; Tinga, Y. P.; Colloids Surf. A (2005), 252, 113–119.

[119] Daniel-David , D.; Pezron , I.; Dalmazzone, C.; Noik, C.; Clausse, D.; Komunjer, L.; Colloids Surf. A (2005), 270–271, 257–262.

[120] Sullivan, A. P.; Kilpatrick, P. K. Ind. Eng. Chem. Res. (2002), 41 (14), 3389–3404.

[121] Spiecker , P. M.; Gawrys , K. L.; Trail , C. B.; Kilpatrick, P. K.; Colloids Surf. A (2003), 220, 9-27.

[122] Sheu, E. Y.; Storm, D. A.; Shields, M. B. Fuel (1995), 74, 1475.

[123] Acevedo, S.; Escobar, G.; Gutierrez, L. B.; Ronaudo, M. A. Fuel 73, 1807 (1995).

[124] Gafonova, O. V.; Yarranton, H. W.; J. Colloid & Interface Sci. (2001), 241, 469-478.

[125] Eley, D. D.; Hey, M. J.; Symonds, J. D. Colloids Surf. (1988), 32, 87–101.

[126] Sjoblom, J.; Mingyuan, H.; Hoiland, H.; Johansen, E. J.; Prog. Colloid Polymer Sci. (1990), 82, 131–139.

[127] Bobra, M.; Fingas M.; Tennyson, E. Chem. Tech. (1992), 22, 236–24.

[128] Bobra, M. "A study of the Formation of Water-in-Oil Emulsions" Canada, (1990).

[129] Bhardwaj, A.; Hartland, S. Ind. Eng. Chem. Res. (1994), 33, 1271-1279.

[130] Tambe, D. E.; Sharma, M. M. J. Colloid Interface Sci. (1993), 157, 244-253.

[131] Raman, S. Manjeri, Polymeric Demulsifiers, U.S. Patents 4120815, (Houston, PA, US), 1978.

[132] Hafiz, A. A.; El-Din, H.M.; Badawi, A.M. J. Colloid Interf. Sci. (2005), 284, 167-175.

[133] Farah, M. A.; Oliveira, R. C.; Caldas, J. N. J. Petroleum Sci. Eng. (2005), 48, 169–184.

[134] Bhardwaj, A.; Hartland, S. Ind. Eng. Chem. Res. (1994), 33, 1271-1279.

[135] Wang, W.; Liu, Y.; Instrument Analysis of Surfactant (Ch), Chemical Induction Industry Publishing Company, Beijing (2003).

[136] Breen, P.J.; Towner, J.W; US Pat. (1999), 5,981,687; C.A., (1999), 131, 1184165.

[137] Frentzel, R. L.; Rua, L.; Pacheco, A. L. US 4460738 (1984), Olin Corp., Invs.: C. A. (1987), 101, 131354.

[138] O'Connor, J. M.; Frentzel, R. L. US 4590255 (1986).

[139] Gaylord, N. G.; Metha, M.; Kumar, V. Org. Coat. Appl. Polym.Sci. Proc. (1982), 46, 87.

[140] Gaylord, N. G.; Metha, M. J. Polym. Sci., Polym. L7tt. Ed. (1982), 20, 481.

[141] Russell, K. E.; Kelusky, E. C. J. Polym. Sci. Part A: Polym. Chem. (1988), 26, 2273.

[142] Minoura, Y.; Ueda, M.; Mizunuma, S.; Oba, M. J. Appl. Polym. Sci. (1969), 13, 1625.

[143] de Vito, G.; Lanzetta, N.; Maglio, G.; Malinconico, M.; Musto, P.; Palumbo, R. J. Polym. Sci. Polym. Chern. Ed. (1984), 22, 1335.

[144] de Roover, B.; Sclavons, M.; Carlier, V.; Devaux, J.; Legras, K.; Momtaz, A. J. Polym. Sci., Part A: Polym. Chern. (1995), 33, 829.

[145] Heinen, W.; Rosenmoller, C. H.; Wenzel, C. B.; de Groot, H. J. M.; Lugtemburg, J.; van Duin, M. Macromolecules (1996), 29, 1151.

[146] Rische, T.; Zschoche, S.; Komber, H. Macromol. Chem. Phys. (1996), 197, 981.

[147] Kozel, T. H.; Kazmierczak, R. T. Ann. Tech. Conf, SOC. Plast. Eng., Tech. Pap. (1991), 37, 1570.

[148] Merck-Bibliothek, Bibliotheks- und Suchsoftware, Opus/ Search (V 1.O), Bruker Analytische MeRtechnik GmbH (1991).

[149] Regel, W.; Schneider, Ch. Makromol. Chem. (1981), 182, 237.

[150] Kellou, M. S.; Jenner, G. Eur. Polym. J. (1976), 12, 883.

[151] Nakayama, Y.; Hayashi, K.; Okamura, S. J. Appl. Polym. Sci. (1974), 18, 3633.

[152] Jacobs, R. L.; Ecke, G. G. J. Org. Chem. (1963), 28, 3036.

[153] Malatesta, V.; Scaiano, J. C. J. Org. Chem. (1982), 47, 1455.

[154] Malatesta, V.; Ingold, K. U. J. Am. Chem. Soc. (1981), 103, 609.

[155] Hamaide, T.; Zicmanis, A.; Monnep, C.; Guyot, A. Polym. Bull. (1994), 33, 133-139.

[156] Buil, C.; Lassau, C.; Charleux, B.; Vairon, J.; Tintiller, P.; Le Hen-Ferrenbach, C. Polym. Bull. (1999), 42, 287–294.

[157] Abele, S.; Graillat, C.; Zicmanis, A.; Guyot, A. Polym. Adv. Technol. (1999), 10, 301-310.

[158] Atta, A. M.; Elsayed, A. M.; Shafy, H. I. J. Appl. Polym. Sci. (2008), 108, 1706–1715.

[159] Davies, J.T.; Proceedings of the second International Congress on Surface Activity, (1957); Vol. 1, London, 426.

[160] Schick, M.J.; Nonionic Surfactants Physical Chemistry, Marcel Dekker (1987) New York, pp.299.

[161] Saito, H.; Shinoda, K.; J. Colloid Interface Sci. (1967), 24, 10.

[162] Wu, J.; Xu, Y.; Dabros, T.; Hamza, H. Colloids and Surfaces A: Physicochem. Eng. Aspects (2005), 252, 79–85.

[163] Nakagawa,T.; Shinoda, K.; New Aspcts In Colloidal Surfactants; Shinoda, K.; Nakagawa, T.; Tamamuushi, B.; Esemura, T. (eds.) Acadimic: New York, 129 (1963).

[164] Anton ,R.E.; Salager, J.L. J. Colloid Interface Sci. (1986), 110, 54.

[165] Atta, A. M. Polym. Int. (1999), 48 (7), 57.

[166] AL Sabagh, A.M. Colloids and surfaces A (1998), 134, 313.

[167] Paszun, D.; Spychaj, T. Ind. Eng. Chem. Res. (1997), 36, 1373.

[168] Tamaki, K. Bull. Chem Soc. (1967), 40, 38.

[169] Connor, P.; Ottewill, R.H. J.Colloid interface Sci. (1971), 37, 642.

[170] Snyder, L. R. J. Phys. Chem. (1968), 72, 489.

[171] Nagarajan R.; Ganesh, K. J. Chem. Phys (1989), 90, 5843.

[172] Al-Sabagh, A.M.; Atta, A.M. J. Chem. Tech. & Biotech. (1999), 74, 1075.

[173] Atta, A. M.; Arndt, K.-F. J. Appl. Polym. Sci. (2002), 86, 1138-1148.

[174] Atta, A. M.; Abdel-Raouf, M. E.; Abdul-Rahiem A. M.; Abdel- Azim, A.A. Progress in Rubber, Plastics & Recycling Technology (2004), 20, No 4, 311.

[175] Atta, A. M.; El-Sockary, M.; AbdelSalam, S. Progress in Rubber, Plastics & Recycling Technology (2007), 23, 241.

[176] Atta, A. M.; Abdel-Rahman, A. A. -H.; Elsaeed, Sh. M.; AbouElfotouh, S.; Hamad, N. A. J. Dispersion Sci. & Tech. (2008), 10, 1484-1495.

[177] Atta, A. M. Polym. Int. (2007), 56, 984-995.

[178] Wertz, D.H. J. Am. Chem. Soc. (1980), 102, 5316.

[179] Rosen, M. J.; Aronson, S. Colloid Surf. (1981), 3, 201.

[180] Rosen, M.J. "Surfactants and interfacial phenomena" John Wiley & Sons, New York (1978).

[181] Halperin, A. Macromolecules (1987), 20, 2943.

[182] Kang, W.L.; Liu, Y.; Qi, B.Y.; Liao, G.Z.; Yang, Z.Y.; Hong, J.C. Colloid. Surf. A (2000), 175, 243–247.

[183] Kang, W.L.; Shan, X.L.; Long, A.H.; Li, J.G. Chem. J. Chinese Univ. (1999), 20, 759–761.

[184] Qiao, J.J.; Zhan, M. Petrol. Process (1999), 15 (2), 1–2.

[185] Foyeke, O.; Opawale, D. J. Burgess; J. Colloid Sci. (1998), 197, 142–150.

[186] Kang, W.; Jing, G.; Zhang, H.; Li, M. Y.; Wu, Z. Colloid Surf. A (2006), 272, 27–31.

[187] Connor, P.; Ottewill, R.H. J. Colloid Interface Sci. (1971), 37, 642.

[188] Davies, J. T.; Rideal, E.K. Interfacial Phenomena, 2nd Ed. Academic Press, New York (1963).

[189] Katalinic, M. Z. Physics (1926), 38, 11.

[190] Shinoda, K.; Carlsson, A.; Lindman, B. Adv. Colloid & Interface Sci. (1996), 64, 253-271.

[191] Buzagh, A.; Rohrsetzer, S. Kolloid ZI (1961), 76, 9.

[192] Shinoda, K.; Yoneyama, T.; Tsutstumi, H. J. Dispersion Sci. Tech. (1980), 1, 1.

[193] Enver, R.P. J. Pharm. Sci. (1976), 65, 517.

[194] Schilinski, H.; Fichtmuller, R.; Tschener, J.; Schmierstoffe Schmierugstech, (1967), 17, 34.

[195] Lissant, K.J. "Demulsification Industerial Applications" Marcel Dekker, New York (1983), 63.

[196] Ahmed, N.S.; Nassar, A.M. J. Polym. Res. (2001), 8, 191.

[197] Kaya, H.; Willner, L.; Allgaier, J.; Stellbrink, J.; Richter, D. Appl. Phys. A: Mat. Sci. & Process. (2002), 74, 499-501.

CHAPTER VII

"SUMMARY"

It is well known that the formation of crude oil water emulsions is responsible for several economic and environmental problems. Demulsification is the breaking of a crude-oil emulsion into oil and water phases. Demulsifiers are class of surfactants used to destabilize emulsions. Recently, there has been increasing interest in the synthesis and characteristics of polymeric surfactants because they probably offer greater opportunities in terms of flexibility, diversity and functionality. The main purpose of this present work is to study the character and demulsification of poly (ethylene glycol)–block–poly (propylene oxide)–block–poly (ethylene glycol) copolymers, setting up the relationship between the molecular structure of PEG–PPO–PEG and the demulsification of crude oil emulsions. In this respect, this this work aims to prepare poly(oxyethylene)-co-poly(oxypropylene) block and graft copolymers to use them as demulsifiers for synthetic crude oil emulsions. New water soluble poly(ethylene glycol)–block–poly (propylene oxide)–block–poly (ethylene glycol) copolymers, PEG-PPO-PEG, were prepared in the normal condition. In this respect, poly(oxyethylene)-co-poly(oxypropylene) polymeric surfactants were prepared by two methods. In the first method PPO was esterified with MA to produce poly (propylene glycol) maleate diester. The produced polymers were esterified with PEG having different molecular weights to produce PEG-PPO-PEG block copolymers. The second method was based on grafting of MA onto PPO followed by esterification with PEG. The first step in this work is to polymerize propylene glycol and propylene oxide using KOH as a catalyst to produce poly (propylene glycol), PPO1, at the reaction temperature 298 K. The reaction time was adjusted to produce PPO having different molecular weights. The molecular weights of the prepared PEG-PPO-PEG block copolymers were determined by GPC technique. The structure of the prepared surfactants was illustrated by IR and ^{1}H-NMR spectra.

The surface activity of the prepared surfactants was determined from the surface and thermodynamic properties of these surfactants. The surface properties were determined by measuring surface and interfacial tensions versus time for a freshly formed surface at four different temperatures (298, 308, 318 and 328 K). The micellization and adsorption of surfactants are based on the critical micelle concentrations (CMC), which was determined by the surface balance method. The CMC values of the prepared polymeric surfactants were determined at 298, 308, 318 and 328K from the change in the slope of the plotted data of surface tension (γ) versus the natural logarithm of the solute concentration. The thermodynamic properties of these surfactants were also determined from the calculation of some parameters such as ΔG, ΔH and ΔS of micillization and adsorption. Studying the surface activity was used to determine the behavior of each surfactant toward micillization and adsorption on its interface.

The evaluation of the prepared poly(oxyethylene)-co-poly(oxypropylene) block and graft copolymers as demulsifiers for water-in-crude oil emulsions was carried out at 318 and 333 K at four different concentrations of demulsifiers (50, 100, 250 and 500 ppm) and at four ratios of crude oil water emulsions o/w (90:10, 80:20, 70:30 and 50:50) respectively. The results showed that the efficiency of the water separation of the prepared demulsifiers reached 100% based on demulsifier chemical compositions and concentrations. It was found that the demulsification times were increased with decreasing the surfactant concentrations from 500 ppm to 100 ppm and with increasing the water cut contents in crude oil emulsions from 10% to 50 %. This behavior can correlated to surface activity, solubility and HLB values of the prepared surfactants. The time of maximum demulsification efficiency was varied from 30 minute up to 300 minutes. The data indicated that the best HLB values for dehydration of the crude oil emulsion are ranged from 11.1 to 12.1. From the obtained experimental results it was found that the prepared demulsifiers showed good demulsification efficiencies towards the different types of water-in-crude oil emulsions.

BIOGRAPHICAL SKETCH

Prof. Dr. Ahmed A. Fadda was born in Cairo, Egypt. He received his B.Sc. 1971 from Faculty of Science, Cairo University, Egypt and M.Sc. 1975 from Faculty of Science, Mansoura University, Mansoura, Egypt. He obtained his Ph.D. in organic chemistry 1981 from Moscow State University, Faculty of Chemistry, USSR, (Ph. D. thesis advisor and topic: Prof. A.N. Kost, "Isomerizaition and recyclization of pyridinium salts to 2-amino biphenyl"). In 1981 he joined the Staff of the Chemistry Department, Faculty of Science, Mansoura University, Egypt. He became Associate Professor at that University in 1986 and was promoted to Professor of organic chemistry in 1991 until now. From 2003-2010, he was the Chairman of the organic chemistry division. He was a visitor professor to North Carolina State University (NCSU) working with Prof. Harold Freeman and Associate Prof. Ahmed El-Shafie in the area of porphyrins, phthalocyanine dyes, photodynamic therapy and photosensitizers. He worked as Professor of organic chemistry at the University of USTHB, Algeria from 1986-1990. He is a member of the Egyptian Chemical Society and the New York Academy of Sciences. In 1994, he nominated by American Biographic Institute (ABI) as one of five hundred leaders of influence (3rd Edition). He is a reviewer for The Permanent Scientific Committee of Organic Chemistry (Associate Professor Careers 1991-present), The Permanent Scientific Committee of Organic Chemistry (Professor Careers 1995-present), Prizes for Outstanding Research, Ain-Shams University (2004, 2006-present), many M.Sc. and Ph.D. theses, large number of international journals and international reviewer for the Academic Scientific Research project's and state awards. He is the principle investigator (PI) of the Egyptian-American joint project. His main research interests are in the field of organic synthesis of heterocyclic compounds of pharmaceutical interest, medicinal chemistry and synthesis of new porphyrins derivatives.

Dr. Rasha Refat Fouad was born in Mansoura, El-Dakahlia, Egypt. She has obtained her B.Sc. (1999) from Faculty of Science, Mansoura University, Egypt and M.Sc. 2005 from Faculty of Science, Menofia University, Egypt. She has obtained her Ph.D. in Organic Chemistry (2011) from Faculty of Science, Mansoura University, Egypt (Ph. D. thesis supervisors: Prof. A. A. Fadda, Prof. A. A. Nassar and Prof. A. M. Atta). She is a member of the Egyptian Chemical Society. She is a reviewer for some of international journals. Her main research interests are in the field of emulsion breakers and organic chemistry.

Made in the USA
Middletown, DE
29 April 2019